Secrets *of* Your
CELLS

Other works by Sondra Barrett

Book

Wine's Hidden Beauty

Video

Super Charge Your Energy: Practical Techniques to Increase Your Energy Based on Integrating the Science of Cellular Energy and Complementary Medicine

Audio

Molecular Messengers of the Heart

Secrets *of* Your CELLS

Discovering Your Body's Inner Intelligence

Sondra Barrett, PhD

Boulder, Colorado

Sounds True, Inc.
Boulder, CO 80306

Published 2013

Cover design by Dean Olson
Book design by Levi Stephen

Printed in the United States of America

Library of Congress Cataloging-in-Publication Data
Barrett, Sondra.
Secrets of your cells : discovering your body's inner intelligence / by Sondra Barrett.
p. cm.
Includes bibliographical references (p. 233) and index.
ISBN 978-1-60407-626-4
1. Cells. 2. Spirituality. I. Title.
QH581.2.B37 2013
571.6--dc23

2011037806

Ebook ISBN 978-1-60407-819-0

10 9 8 7 6 5 4 3 2 1

To Alvaro and Paulo, who wanted to see
and know what was inside, and whose lives
helped illuminate my sacred path

and

to my children, Ted and Heather,
who illuminate my heart and soul,
filling me with love

The best way I have to honor God

is to understand the secrets of Nature.

—DR. MICHIO KAKU theoretical physicist

Contents

Illustrations

Figures

Color Plates

Preface

Everyone who is seriously involved in the pursuit of science becomes convinced that a spirit is manifest in the laws of the universe—a spirit vastly superior to that of man.

—ALBERT EINSTEIN

The voyage that led me to bridge the realm of the biological cell with the realm of the soul was a trek that required the repeated breakdown of my ego and beliefs. It began more than forty years ago when I obtained a PhD in biochemistry. Back then, I was intrigued by the chemicals of life. They were identifiable, objective, and quantifiable; plus, I believed that if we could locate chemical abnormalities, we could fix anything in the human body. Later I pursued immunology and hematology, which took me into the cellular universe.

As I explored cells under the microscope, my experience became more than cerebral—I was *enchanted* by what I was seeing. I began photographing that magical microscopic world of living human cells (see plate 1 in the color insert for my first photograph).

While the world of cells captivated me in a mysterious way I didn't yet fully understand, I was very much in an intellectual mode. I thought that unless something could be measured and proven, it wasn't real. It was an illusion—if not a delusion. In my mind, analysis and statistics told the true story; there could be no ambivalence about the conclusions. However, the more experience I gained through my research, the more cracks appeared in my rigidly held convictions. I saw people die who "shouldn't have" according to their biological measurements. I met children with aggressive leukemia cells (cancer of the blood) who contradicted the predicted course of their disease and were living much longer than they "should have." People didn't fit so nicely into statistical categories and prognostications. They couldn't be easily measured like a test tube of defined chemicals.

One eighty-four-year-old woman, Marjorie, with acute leukemia stopped responding to chemotherapy. Her doctors gave her just a few months to live. Marjorie, however, had other plans: a grandchild was graduating from college and her sixtieth wedding anniversary was coming up. She needed to stay alive for these events. And she did—without chemo. In fact, she remained with the important people in her life for two more years.

This made no sense to me. I was stymied. What kept her alive? My shell opened just enough to realize that healing, life, and death could not be placed simply into an analytical framework, and I could no longer depend solely on the comfort of scientific measurement and predictability.

Then the father of a boy with leukemia asked me to photograph his son's cancer cells to aid in a visualization practice. The boy would imagine each healing cell as a kind of Pac-Man who sought out and destroyed the cancer cells just as it ate pellets in the arcade game. This was a time when visualization and imagery were found only at the fringes of mainstream medicine, but the concept took hold in my imagination. I thought that if kids could see what healthy cells looked like—and those cells were bigger and stronger than the cancer cells—they might be able to use their minds to heal their bodies. Sometimes I would suggest that they imagine their cancer cells as dust and their healing system a vacuum cleaner.

Soon I began giving weekly "inner space" slideshows at the clinic. I simply pointed out healthy and abnormal cells and what the molecules were—with no further explanation. It became clear to me that to the children it didn't matter what these images meant to a scientist. *There was something inherent in them that people of all ages enjoyed and that many experienced as transformative.*

I formed a special bond with a five-year-old boy named Alvaro, who asked to see the slides over and over again. I sometimes invited him and his sister to share the weekend with me and my children, and we would sit together and draw cells or go to the park. Then suddenly, Alvaro began to go downhill after being in remission for more than a year. His speech faltered and walking became difficult. What could I do for him now?

I remembered learning a Gestalt psychology strategy in my own therapy sessions to express difficult feelings (hitting a pillow, yelling) and was inspired to try it with Alvaro. I asked if something was upsetting him and was surprised when he immediately answered that he was really angry at his stepfather. He told me he believed his real dad had been pushed out of the home by this man. In my innocence—untrained as a Gestalt therapist—I told him to show how angry he was by pounding on the sofa. He didn't hesitate and beat on the sofa pillows for quite some time.

A few days later, Alvaro began to improve. While it could have been the new meds he had been given that caused the change, to me, it seemed a miracle. I could no longer be *certain* that it was *only* the medication that turned his disease process around. This marked another major turning point in my thinking and beliefs about medicine. It was at this time, fearing Alvaro's death, that I sought the help of a clinical psychologist. During my first therapy appointment, the psychologist lit some sage to purify the space, something I had never experienced before. I immediately felt more clarity and relaxation than I had in a long time. It was clear that this man was more than meets the eye—he was, in fact, a shaman. He brought me to a deep place of trust, and I knew I needed his guidance in the big issues I was struggling with, which my scientific training offered no context for. I saw him as a therapist for several years before I committed to a year-long shamanic apprenticeship.

My work with this shaman was a true turning point. He helped me reframe my concepts about what healing is and the mental and spiritual dimensions that foster it. By going deep into my own healing journey, I began to explore my own role as a healer. During that potent apprenticeship year, I made my lifelong aspiration to bridge science and spirit in the healing process.

In the lab, my most significant clinical research was to explore how to recognize cellular characteristics that could lead to more accurate diagnoses and successful treatments of leukemia. Using the microscope to distinguish cellular identities and behaviors, I learned that white blood cells change shape and form as they grow and mature. As their forms changed, so did what they were able to do. Compared to the orderly and regular forms of normal cells, the expression of leukemia cells was

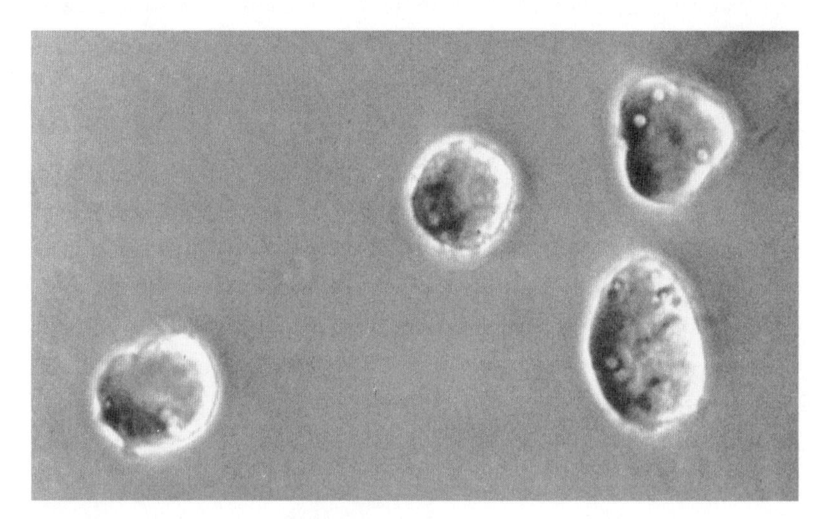

Figure P.1 Living human myeloid leukemia cells

chaotic. Here is a photomicrograph of myeloid leukemia cells, each having a different and abnormal shape (see figure P.1) when compared with the regular shape of normal myeloid cells.

Years of research on the part of myself and my team led to successful results: more definitive diagnoses of the different forms of these most deadly acute leukemias. Yet while this was a tremendous breakthrough, I felt like a failure. Had I asked the wrong question by focusing on diagnosis? At the time, there were no new treatment options for these diseases—no one was going to get better because of my work. I considered leaving laboratory research.

And yet, my work examining living cells under the microscope had affected me profoundly. During one very early experiment, I watched living human white blood cells detect tiny, inert plastic beads. Instantly the cells went into action. They slithered and morphed as they moved rapidly toward the plastic spheres in an attempt to eliminate their intrusion. Here was a mystery, unfolding before my eyes. Could white blood cells possibly be so smart by accident? Or by a whim of biology?

Influenced by my shamanic studies, as I witnessed living cells' heroic efforts to defend against danger, I began to see them as more than programmed tissue. They were holy. They were evidence of God's handiwork, of a divine design. I now accepted that the invisible world was more than the cells and molecules I had studied in books and thought I knew; this world encompassed spirit, wonder, and the soul.

Not long after my dear little friend Alvaro died, I received a diagnosis of hepatitis. Each morning at the hospital, I had been the "balloon lady," handing out balloons to the kids to blow up and then draw on. If they couldn't blow them up full enough, they'd give them back to me—gleaming with saliva—to finish. Who knew then that this could pose a risk to my health? These were the days when few precautions were taken with blood (which I handled all day in my work) and saliva.

Puzzled by the worsening course of hepatitis, my doctor told me, wisely or unwisely, there was a distinct possibility I could go into a coma, and if I did, I could die within twenty-four hours. This frightening prognosis changed the course of my life. I reasoned that if I were to die young, I'd better reset my priorities. I spent more time with my children and moved out of San Francisco to the beach. There, I met people who were on an alternative healing path, far different from mainstream academic medicine. Healers and yogis, organic gardeners, holistic doctors, and prolific poets visited my new home to share ideas. From them I managed to learn more about healing—and how to save my own life.

For the first time, I experienced bodywork, acupuncture, and fasting. I sought to cleanse my body of the virus and my mind of disappointments. Through my own personal experience, I became convinced of the usefulness of "alternative" healing strategies, even when their efficacy could not yet be proven by Western science. I came to accept that more than the body needs healing—mind, emotions, and spirit each play their part. I saw that emotional and professional disappointments could have contributed to my inner environment being more susceptible to illness.

In turn, my new neighbor friends began asking me questions to help them understand biology and chemistry. I began teaching for the first time, which forced me to deepen my own scientific knowledge. I had to simplify concepts in order to explain them. In other words, I had to understand science better myself.

One outrageous neighbor, the very persistent "Princess of Argisle," as we affectionately called her, discovered my penchant for photographing everything under the microscope and urged me to take pictures of minerals related to astrology. *Astrology?* I had opened my mind to many new realms of knowledge, but astrology still struck me as nonsense. But because the minerals she was asking about are a part of human cells as well, and because I knew I could use the resulting images to teach children about their bodies, I finally agreed.

When I saw that the photographs of the twelve mineral salts revealed only four distinctive shapes, my curiosity was piqued. There seemed to actually be a connection between these molecular patterns and the symbols of astrology. With a little research I discovered that the four shapes

I saw corresponded to the four astrological elements of earth, water, air, and fire. See four of the mineral photographs—representing these four elements—in the color insert. Further, I learned that this kind of correspondence between physical form and symbolic meaning had roots in ancient medicine, language, and Jungian psychology.

I remained a skeptic, yet I was intrigued. Could modern microscopic patterns possibly align with old occult precepts and ancient wisdom? Was I uncovering another example of "as above, so below"? Had I discovered metaphysical meaning in our molecules beyond their chemical significance? I came to appreciate that *our cells and molecules are divinely designed following sacred universal laws of nature.*

> *Detecting designs and patterns where no designs and patterns were previously apparent can produce tremors of faith. . . . As far as contemporary science can tell, nearly everything about the universe—its knack for self-organization; its fine-tuned potency to bring about galaxies, life, consciousness; its sheer existence—is vastly improbable. This would seem to suggest that we are here because of a deliberate supernatural design.*
>
> —HERBERT BENSON, MD *Timeless Healing*

I was nowhere near the lab or microscope when another light bulb came on. Instead, I was in the Southwest photographing indigenous cave paintings. Patterns and connections began to emerge, and I soon interpreted a thousand-year-old Native American medicine wheel as a stylized version of a cell. See plate 4 in the color insert to see a medicine wheel pictograph at the Palakti ruins. It has the same construction as a cell: its center circle is like the core nucleus; the lines on the outer surrounding circle could represent cell receptors and markers of identity. The four sets of three spokes point to the four directions, a central concept of Native American cosmology; the cell, too, has triad structures that point to the cell's direction. And so it was that I leaped to the possibility that the painting could represent more than we have thought.

Figure P.2 "DNA" pictograph at Palakti Ruins

In that same cave, I saw another pictograph that could easily be interpreted as a drawing of DNA (see figure P.2)—that is, if you were thinking about cells.

After this experience, I was off, exploring how our human microenvironment might be reflected in other ancient symbols. Because anthropologists can only speculate what these ancient figures mean, in my mind they could easily have sprung from the imagination or inner vision. Shamans and indigenous people, spiritual seekers, and students of dreams all bring descriptions of images they have seen in the dreamtime into everyday reality. Could these forms just as easily come from imagination or the "seeing" of our inner world as from something visible to the eye?

By the time I contemplated the cave paintings, I had experienced powerful shamanic imagery in pursuit of healing that came through altered states—mainly deep meditation—so I knew it was possible for information to emerge through different states of consciousness. This is not to say that ancient people named what they saw *cell* or *DNA;* it took scientists hundreds of years of study to name and show us these realities.

Nonetheless, there is considerable evidence that inner vision can lead to outer manifestation.

This theme that the architecture of our molecules and cells provides an underlying framework for spiritual teaching and sacred art pervades my journey. Sometimes I call this *cellular anthropology.* Anthropology being the study of human cultures, cellular anthropology, then, is the exploration of how our cellular architecture has influenced human culture. If we look at how ancient traditions have added to modern knowledge, the invisible world could actually have been part of ancient knowledge. For example, for centuries, people have used the form of the mandala to help center themselves, to access the sacred. The image created by Dr. Robert Langridge, shown in plate 5 in the color insert, may appear to be an artist's mandala, yet it's actually a product of modern technology, a computer graphic of DNA, looking at DNA from the top of the molecule. Is it a mandala or a molecule? Art or science? Ancient or modern?

The roots of knowledge come from many spheres. I have come to understand myself as a "code finder," whose unexpected path is to uncover and make visible secret messages hidden in the very architectures of life that build us, as well as our holy traditions. I see sacred geometry in our molecules and I see a story of creation as told by our molecules. While it may seem as though cells and molecules are the focus of my work, in fact I sought to offer the whole picture of how mind and body engage and influence our invisible terrain. By teaching the new field of body-mind medicine, or psychoneuroimmunology (PNI), I became convinced of how our healing systems are all connected. My students asked which practices worked best for stress reduction and body-mind healing. For example, does imagery really work? I had to find out, so I began seeing clients in a clinic in Marin County.

Soon, I was leading healing groups, not just teaching classes. Group work takes us into the realm of psychology, not my expertise or experience—I was a scientist in the realm of the physical, not trained or "qualified" to help with the human condition. Nonetheless, over the years, I developed what I called *psychoeducational groups* for adults with cancer, autoimmune illnesses, and heart disease. I taught the biology of what we knew about these problems and then offered practical solutions for dealing with the disease and its related stress. We practiced imagery, qigong, sound (using chant and toning), and numerous relaxation strategies from my "medicine bag of tricks." As one of the early teachers of PNI to the lay public, I was invited by an organization to deliver continuing education programs for health professionals. I traveled around the country teaching about the immune network, energy, and stress management.

Two weeks after the World Trade Center towers in New York collapsed, the Pentagon was hit, and a terrorist-piloted plane crashed into a field in Pennsylvania, I flew toward these same places. My mission, scheduled months before, was to teach stress-reduction and energy-management practices to health professionals—to people who had suddenly found themselves on the front lines.

When I arrived, the whole scene implied war. I was terrified, and I was supposed to be teaching health professionals how to balance their energy—now, in the midst of a catastrophe. How could I possibly help the nurses and psychologists working on the front lines renew themselves and the people they were trying to help, people who were experiencing a depth of fear heretofore unimagined?

I prayed for guidance, and an answer emerged: *Bring your shamanic spiritual wisdom to these people.* I hesitated. I was supposed to be giving them science for continuing education credits. More guidance came: *Give them both—words that support their intellect and internal skills to tap into the heart of their spiritual intelligence.* As never before, I was asked to embrace all that I had ever learned about the healing of mind, body, and spirit, and about the bridge between science and soul. To these large groups I taught a simple health-enhancing qigong exercise and a guided meditation that had helped me in the most stressful times—both of which are shared in this book.

Under such unfavorable conditions (strangers in large unfamiliar hotel rooms), and after the psyche-shattering experience we had all been through, I didn't expect people to get deep into their feelings or be willing to share. I was in for a big surprise. Several people said it was first time they had been able to cry since the tragedy—none of these heroes had been able to let go until then. That trial by fire convinced me that I had more to teach than science. I could bring practical applications to the heart of healing.

The book you are holding contains the harvest of my long journey to join the worlds of science and spirit. I have written it because I know I have a unique perspective to offer about cells and molecules in relationship to ancient wisdom. I want to offer this new view to people who are hungry for spiritual connection and for knowledge about how best to care for themselves. And I want to demonstrate the sacred within, the bridge between science and healing. I see it as life's operating instructions: lessons from our cells.

The scientific doorway opens onto exciting discoveries. The realm of the spirit reveals deep truths. In this book, we will explore the courtship between science and spirituality and discover practical ways of healing while transforming and infusing mind and molecules with the sacred. The cell itself will be our guide.

Introduction

Man is a colony of cells in action. It is the cells which achieve, through him, what he has the illusion of accomplishing himself. It is the cells which create and maintain in us . . . our will to survive, to search and experiment.

—ALBERT CLAUDE 1974 Nobel laureate in medicine

You are about to embark on an extraordinary voyage. As you enter the chapters of this book, you will don the identity of a new kind of adventurer: a *cytonaut,* a "sailor inside the cell." Like Alice down the rabbit hole suddenly grown small, you will find yourself in a mysterious new world, and there you will explore the structure and workings of the trillions of tiny cells that compose *you.* You will come to understand that the living cell, the perfect container for the divine spark of life, contains more than scientists may be willing to admit—more than nucleus and membrane; receptors and genetic markers; fluid, flexing strings, and tubes. You will find that it also holds important lessons about how to live a fuller, healthier life. And you will encounter the compelling notion that the shapes and movements of the cell, visible to the eye only with the aid of the microscope, have been intuited for millennia by seers and shamans and are present in ancient art found around the world. *Secrets of Your Cells* opens the door to cellular intelligence and ancient wisdom, the magic and majesty that dwell within you.

About the Cell

You will have ample opportunity to investigate the nature of the cell throughout this book. By way of introduction, let me just say here that our cells are our oldest living ancestors, shared by all of life since its creation. We all possess the same building blocks, molecules, and biochemical principles. Written in the biography of the cell are the mysteries of life, growth, and transformation. At every moment of every day, our cells orchestrate millions of molecular symphonies, guided by cellular intelligence in a delicately designed system of checks and balances, push and pull, collaboration and communication. Fundamental to cellular functioning is the *molecular embrace,* in which elements of the cell fit together like hand in glove to realize their combined destinies; connection is, in itself, a building block of life.

When we examine the life of the cell, we are witnesses to sheer genius. In my view, based on much experience and study, there is clearly an intelligence at work. When you have completed the journey at hand and turn the page on the final chapter, concluding your own investigation into the intricacies and dynamics of cellular life for now, you, too, may find that you have adopted this view.

Cells and the Sacred

This book speaks to two dimensions of our human experience: scientific investigation and spiritual exploration—here we bridge science and the sacred. When we mine for scientific facts, we are engaging our intellectual brain, the one that wants data, analysis, proofs, and measurements. The scientist wants to know why and how. If you also want to know the physical, perceivable whys and hows of the cell's functioning, you have come to the right place; I have allowed the scientist in me full rein.

The sacred and spiritual experience is said to have no place in science, yet to fully know and appreciate life and our place in it, I believe both dimensions need to be present. Our inner, intuitive, natural knowing, which I sometimes call the feminine side of science, looks at the whole of experience rather than only its measurable components. Cells can be appreciated for their philosophical teachings as well as their physiological abilities, and this book speaks to both. Our cells are little crucibles of measurable, discernible biochemical interaction that also carry the seeds of divinity. There is poetry written in the scientific alphabet of life; molecular scribes have much to say. So I ask you to open your mind and your heart as you investigate the cellular world for yourself, for the cell holds deeper truths than can be found in its wondrous scientific realities. We are a constellation of trillions of individual energy holders—soul keepers, I believe—that carry the wisdom of the ages and keys to the Mysteries.

A Playbook and Guide

I have designed this book to be both a playbook and a guide to the cellular universe. It will encourage you to engage with your cells in very practical ways to invigorate your body and ignite your imagination. I want you to fall in love and enjoy an intimate relationship with your cells.

Cranky? Your cells receive the tense messages you are sending them and tighten up. Feeling relaxed and peaceful? So, too, are your cells, moving with ease and efficiency. Our choices influence their life experience and ours; cells respond to what we give them. When we bring them fresh air, they are able to produce energy more efficiently. If we nourish them with love, laughter, and music, pleasure-inducing endorphins flood our being with happiness. When we worry, our internal pharmacy bombards them with stress hormones that can damage them—and us. I have sprinkled throughout the book information and explorations to help you learn how to treat your trillions of tiny building blocks—and yourself—in nurturing and life-affirming ways.

Among the explorations you will find are what I call *body prayers,* a term I have borrowed from a friend and colleague who once orchestrated for me a wonderful ritual dance she called a body prayer. A body prayer is a sacred movement, one in which the soul expresses itself. Many of the body prayers contained here are qigong exercises; these are adapted from ancient Taoist practices and combine mental focus, breathing, meditation, movement, and imagery. They are meant to get you moving, and if you perform them with intention, they will energize your body and transform your consciousness. They will help you create daily awareness that supports change, inspires commitment, and imparts sacred knowing.

Moving through the Book

Each chapter will take you into a specific feature of your cells while exploring its spiritual potential and offering activities to engage your mind and body. As you move through the book, your body of scientific knowledge will build, and you will deepen your experience of your own cellular life.

In chapter 1, "Sanctuary–Embrace," we explore *sanctuary* as we discover our cells' creation story, consider the cell as a sanctuary and container for life, and learn about the features of our cellular membrane. Chapter 2, "I AM–Recognize," takes us into the cell's recognition capabilities—discerning self from other, carrying markers of identity, and wielding the wonderful complexity of the immune response. The larger lesson about self found here is present in ancient scripture: I AM THAT I AM. And in this chapter, we learn to make our cells hum. In chapter 3, "Receptivity–Listen," we travel from the "I" to the "we"—cellular communication and listening. We learn about the nature of our cell membrane receptors and how our cells tune in to the vast array of information signals they receive.

We find that our cells always live *in the now,* and we are encouraged to follow their lead.

In chapter 4, "The Fabric of Life–Choose," the concept of the "brain" of the cell, so brilliantly described by pioneering scientist Dr. Bruce Lipton, is expanded to include an even more embracing intelligence contained in our cells: the cytoskeleton. We see that the *scaffolding and fabric* of the cytoskeleton are the likely anatomic location where energy healing takes place and consciousness resides. There are also lessons for us here about attachments and letting go. Chapter 5, "Energy–Sustain," is an exploration of energy: how our cells use, make, and conserve it, and how we, the larger organism, can sustain and maintain it. Chapter 6, "Purpose–Create," moves us deep into our spiraling DNA molecules and genetic expression and takes a look at what can go wrong with our DNA as well as our self-correcting abilities. At this point, we will also begin to look at metaphysical symbols that echo the rhythms and patterns found within the cell.

Chapter 7, "Memory–Learn," brings us to an exploration of cellular resonance and holographic memory plus the role our senses play in remembering. We experience how to create cellular networks to reinforce and strengthen learning and memory and investigate how to create new habits and break old ones. In chapter 8, "Wisdom Keepers–Reflect," we leap from inside the cell to see its characteristics reflected in the myths and symbols of ancient civilizations. Finally, in chapter 9, "Connection–Cell-ebrate," we recall our entire voyage, mine it for some of its key lessons, and give gratitude for all we have shared.

Here I have given you a thumbnail sketch of the terrain you are about to travel. I wish you a rewarding voyage full of wonder, inspiration, and discovery.

And now, ladies and gentlemen, I give you your cells . . .

Chapter 1

Sanctuary–Embrace

Each one of us carries a fifteen billion year existence in us, so that when we encounter one another we ought to be awed by the experience. And when we encounter ourselves . . . every hydrogen atom in our bodies has been in existence for fourteen billion years—imagine how many stories they have to tell us.

—MATTHEW FOX *One River, Many Wells*

Do you ever think about how life began or how the earth was created? I've read myths about the origins of life and reviewed scientific experiments that seek to determine how it all began. But it wasn't until a little girl asked me where her brother would go when he died that I thought more deeply about what I knew about creation, life, and death. After that conversation, I created a collage to reflect my view of the question, embracing both dimensions that I address in this book: science and wonder, molecules and mystery. We will consider these topics at the beginning of this chapter.

Then we will move past these questions and examine life as we now find it. We will look at how our molecules form a container for life, and how our cells' inherent intelligence generates the know-how to enable us to survive and thrive. And because this book straddles two worlds—science and the sacred—in learning about the ways of our cells, and the molecules of which they are made, we will have opportunities for both personal and spiritual discovery.

In this chapter we discover *sanctuary.*

Origins: Myths of Creation

"In the beginning . . ." This is how many stories of life's origins start. But what was the beginning? Before there was life, what was there? Emptiness, the abyss, a void? Rocks and water? Is physical creation purely a random accident of nature, or was it divinely guided? These have been questions throughout the ages.

Myths and stories give us a way to think about life and our place in the universe, and every culture has a treasured myth that explains symbolically how life began. Most religious and spiritual traditions also tell tales of creation and how we humans came into existence. The Old Testament underlying Judeo-Christian beliefs actually offers two creation stories: the seven days of creation and the Adam and Eve story. The New Testament adds an additional idea that "in the beginning was the word" and that the word was from God: sound created the universe.

Sound is a primal force that when studied through the science of cymatics can be shown to organize physical matter.[1] Cymatic experiments using varying sound vibrations have revealed an astonishing array of forms developing in amorphous sand particles or fluid water. If sound vibrations can demonstrably cause the formation of shapes in matter, it is not a great leap to imagine that sound had a hand in shaping life. In fact, according to Hindu, Buddhist, and Sufi scriptures, it was indeed sound and an even more subtle energy—the vibration of thought—that created the universe.

Modern science's myth of creation is the big bang, a colossal explosion that started it all. Here again we find sound at the center of the story of our beginnings.

I take this idea into present-day reality by recognizing the ways in which sound—including music and the "noise" of our thoughts and beliefs—shapes our reality, our continuing life moment by moment.

In the scientific creation myth, our beginnings occurred in a vacuum in which no sound could be heard, and of course hearing creatures did not yet exist. Then the universe burst into existence. Renowned musicologist and drummer Mickey Hart puts it this way: "Fifteen or twenty billion years ago the blank page of the universe exploded and the beat began."[2] What emerged from the cosmic soup of neutrinos, photons, quarks, and strings were rhythmic vibrations that keyed the formation of galaxies, solar systems, planets, us. The vibrations of the big bang reverberated across space, organizing elemental atoms into simple gases and then increasingly more complex molecules and life. Though this origin story encompasses scientific facts from the disciplines of astronomy, quantum physics, chemistry, and biology, ultimately it probably cannot be proven, so like its religious counterparts it lies in the realm of mythic conjecture colored by cultural beliefs.

> *Dying stars are intimately related to our existence as living organisms on earth. All elements necessary for life—carbon, nitrogen, oxygen, iron, etc.—are manufactured in the nuclear furnaces in the interiors of stars. It is in the dying of stars and their release of these components that life has its birth.*
>
> —MICHAEL DENTON, MD, PHD *Nature's Destiny*

I used to think that people who said we came from the stars were simply exercising poetic license. Yet in our evolutionary trails and tales, the heavens and stars, including our own sun, existed long before living things. In fact, we now know that our earth contains minerals that were set free by the expansion and explosion of stars. The combination of these elements created both stable and unstable forms. Those that survived long enough joined with others to build more and more complex structures—and we, perhaps the most complex of all, are the intricate product of these molecular constructs. We literally embody the scientific creation myth.

An Alchemical Creation Tale: Molecular Embrace

My interpretation of how life began on this planet is this: Simple molecules sloshed and mingled in hot, churning waters. Carbon and hydrogen atoms, our earliest molecular ancestors, formed chains—you can envision them as pearls, or beads that snap together. Molecular strands organized themselves into sheets and wiggly shapes. Then, somewhere along the way to cellular creation, a major disturbance occurred: a lightning bolt, a falling meteorite, a vast explosion, a shockwave. Dramatic and momentous events brought the hydrocarbon strings closer together; the floating molecules found one another. They swirled, gathered together, and embraced while clinging to the salty waters in which they were floating. And their merging formed *a container for life.* Life needs a container in which to exist!

Once I captured this scene clearly in my mind, I wrote this poem. I haven't changed a word of it, even after learning that University of California professor Dr. David Deamer, who has been researching our origins for thirty years, believes that the gathering of molecules occurred in warm tide pools rather than hot, turbulent seas.

> *Once upon a time, before there was life,*
>
> *The world was hot churning waters and floating molecules.*
>
> *A disturbance in the air! Lightning! Fire! A Big Bang!*
>
> *Suddenly molecules embrace, creating*
> *sanctuary for tiny drops of the sea.*
>
> *Over eons, the invisible, shimmering, oil-*
> *covered droplet slowly emerges into a cell filled*
> *with promise and the divine spark of life.*

According to the Jesuit priest–geopaleontologist Pierre Teilhard de Chardin in his book *The Phenomenon of Man,* it is because of the very nature and ability of our molecules to embrace that we, too, embrace one another.[3] Though it may be hard to imagine molecules in a romantic embrace as if they were conscious, intelligent, or loving, their molecular wisdom creates all the right forms through embrace. Molecules merge to form a vessel with a resilient, flexible surface that protects, defends, and defines the cell self.

Life Needs a Place—The Cell as Sanctuary

Without a container . . . there can be no life.

—CARL ZIMMER "First Cell," *Discover* magazine

Part of the definition of life is that it is in a place.

—DAVID DEAMER Interviewed in "First Cell"

Whatever ideas we hold about how life began, there can be no doubt that we humans are a molecular sea. We are constructed from a vast collection of chemicals; yet chemicals aren't enough for life. If we mixed together all the essential ingredients—sugars, fats, amino acids, DNA, RNA, minerals, and vitamins—we still wouldn't have life. The energetic spark, our life force and essential spirit, is different from our chemistry. Only when the divine spark *finds sanctuary in form* can life begin. The residential sanctuary for life is the cell.

About three hundred years ago, English scientist Robert Hooke, looking at a slice of cork under a microscope, saw delineated spaces and porous structures and named them cells. Living cells, however, differ significantly from porous, dead cork. Hooke did not view the structures he saw as the basic units of life; rather, he saw them as containers that held "living juice."

Consider this for a moment: a cell is a place where life is captured and vital ingredients are stored and protected—it is a sanctuary for life. Within each of us, therefore, are 100 trillion cells—100 trillion sanctuaries.

Definitions

Cell, from the Latin *cella:* A storeroom or chamber. This is one definition of a cell. Another is a small room in a monastery or convent. There is also the prison cell, a space you are not encouraged to call home.

Sanctuary, from the Latin *sanctuarium, sanctus:* Holy. A sanctuary is a place where one reflects on and takes time to honor the sacred. It is both a place for microscopic life to exist—the cell—and a place for the human being to reflect on what is holy. Both kinds of sanctuaries, the physical and the sacred, coexist within us, and when we understand this and see the connection, the idea can be life changing—it was for me. Having gained this insight about sanctuary, I began seeing my cells—and hence myself—as sacred. I began to think of my cells as an integral part of my life and remembered to care for them. I could decide, for example, to consciously engage in the simple act of taking my cells for a walk, thereby energizing them and consequently all of me. In remembering to nurture my cells, I became moved to transform unhealthy patterns and recognize that my cells and I are in life together.

Sacred: Consecrated. Devoted or dedicated to a deity or religious purpose. Holy. Entitled to reverence and respect. Synonyms: Revered, cherished, divine, numinous, blessed, immune, godly, spiritual.

Reflection

Take a moment to reflect on how you are a sanctuary. What is it that creates in you a sense of peace and safety? Does one part of you feel more like a sanctuary than another part?

Exploration

From Inner to Outer—Your Sanctuary

After reading the following instructions, close your eyes and take an imaginary journey to enjoy your personal, lifelong sanctuary.

Set aside five to ten minutes in a place where you will not be disturbed. Notice your breathing. Breathe into what you consider your center. Then breathe to your outer edges. Feel the connection between your innermost center and your outer boundaries. With each breath, you are filled with life. With each breath, feel the sanctuary of your cellular universe. Allow yourself to experience the deep connection from your center outward: this whole body is your sanctuary. You can come here at a moment's notice—no reservations are necessary. Stay in this embrace of your breath and cells until you are ready to leave. Then, when you are ready, open your eyes, shake out your hands, and smile to your cells. Thank them for crafting the ingenious container that is you.

◇◇

Creating Sanctuary: The Architecture of the Cell Membrane

To gain access to the mysteries of life, we begin by exploring the workings of our cellular container. The outside surface of each human and animal cellular sanctuary is called the *plasma* or *cell membrane* (see figure 1.1). It is flexible, malleable. (Plant cells, by contrast, have more rigid walls, not membranes.) The plasma membrane, constructed from self-organizing chains of hydrocarbon molecules (fat) and generously "seasoned" with protein, forms the boundary where the cell meets the world in which it lives. These fats have unique qualities: one end of the molecule loves water, while the other end repels water and is called *hydrophobic.* Two layers of fat form the membrane, with the hydrophobic "tails" connecting together inside the membrane and the water-loving "heads" facing out into the watery external environment and in toward the "innards" of the cell. The highly intelligent and resilient surface of this flexible, fatty covering has many functions:

- It protects and insulates what is inside.
- It provides a discriminating, fluid, semipermeable barrier.
- It evaluates what is allowed to come in and go out.
- It carries passwords of identity and receiving sites for essential information.

In general, a cell's plasma membrane becomes more flexible as the cell matures. It also becomes better able to recognize other cells, respond to the environment, move, and change shape.

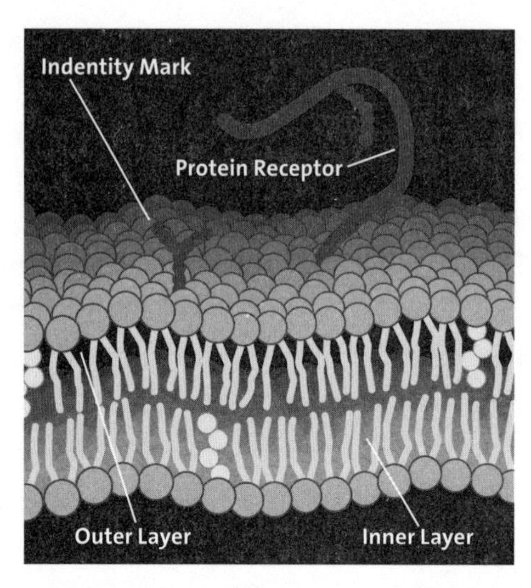

Figure 1.1 Cell container

Magnified through the microscope, our cells appear quite solid. Yet when we peer deeper into the plasma membrane, we discover that this protective, mayonnaise-like covering is rather open, with its component molecules moving within it. The healthy functioning of our cells depends on the fluidity of our cell membranes, and this is partially influenced by what we eat.

Nourishing Our Cell Membranes

The primary chemical components of our membranes are fats, proteins, and cholesterol. The physical properties of fats and cholesterol control the membrane's shape, function, and fluidity. A current theory is that excess dietary trans fats, saturated fats, and cholesterol make the cell membranes more rigid. When the long, straight chains of saturated or trans fats are incorporated into the cellular membrane, they render it less flexible. Meat, butter, and animal products are sources of saturated fats; trans fats are found in hydrogenated oils and margarines. Unsaturated fats, by contrast, have kinks and bends that provide space within the membranes, allowing for more movement and flexibility. Sources of unsaturated fats include olive oil, nuts, and other vegetable oils. The other component of our cell membranes, cholesterol, whether made by our cells or taken in through our diet, can also influence how rigid the membrane is.

Why are cell membrane fluidity and flexibility so important? Because they determine how well nutrients can enter the cell, how readily the cell's receptors can respond to information, and in the case of immune cells, how efficiently they can eliminate pathogens (disease-causing agents).

Reflection

How fluid is your diet? Do you enjoy foods rich in unsaturated fats, like avocados, nuts, and fish? Are you eating too many packaged foods full of trans fats? Consider choosing foods that nurture your cells, your sanctuaries of life.

Proteins, the third major component of plasma membranes, are large molecules and have particular orientations within and across the membrane. Some proteins straddle the double-layered fatty membrane, reaching inside the cells from the outside, while others float on the surface. Fats and cholesterol in the membrane are the medium through which proteins move and function. The proteins themselves are the action molecules, conferring ability and identity to the cell. Proteins serve as both receptive antennae for information and distinguishing markers for identity. You will learn more about both in upcoming chapters.

The cell membrane carries on its surface the cell's identification "passwords" as well as its listening devices for rapid communication.

Exploration

Embodying the Container: Self-Discovery

By learning to embody the cell, we can gain practical and profound wisdom about its nature. Let's face it; though we are constructed from trillions of cells, to most of us, cells are an abstract concept—nearly fictional. This exercise helps make cells more real and tangible. One of the first things I do with adult students in my Cells and the Sacred workshops is encourage them to "become a cell container." Here's how:

Gather together at least four people and sit on the floor in a circle. Face away from each other with your backs to the center, shoulders touching. Together you are forming a container, protecting the sacred space inside. Close your eyes and become aware of what you hear, sense, and feel. After a few minutes, open your eyes and again take in what you see, hear, or sense from your surroundings.

After several minutes of facing outward, turn inward to the center of your "cell" and close your eyes again. Listen with attention for a few minutes and then open your eyes, gazing into the space in the center. With the rest of your "cell," discuss your experience for ten to twenty minutes.

What did you learn about yourself and your cells? What did you notice when facing outside? Was that different from facing the center?

Here is how two workshop participants experienced the exercise.

> Acting like a cell, we were asked to simply experience what we felt. Facing out was surprising for me. I felt protective of what was inside the circle, like I was standing guard. At the same time, I was receptive to what was before me. I felt open and expansive. —MP

> Our cell-making exercise was a lesson in self-organizing systems and an experience in the meaning of boundaries. With eyes closed, we turned to face the world beyond our newly formed cell. Some members felt protective of the group as they heard approaching noises. Others felt the fear of responsibility for holding together their part. We each made a different connection with the external world depending on our unique perspective. I became intrigued with the ability of our cells to receive information. —JM

A Cell's Way of Life

Now that you have experienced the cell as a container and have a more tangible sense of it, consider this question: what makes it alive? In achieving and sustaining life, nature uses a universal set of building blocks: carbon atoms, water molecules, and genetic codes that hold a vast array of life stories.

The cell is the smallest functional unit of life—and believe it or not, scientists don't always agree on the definition of life. We do know that all life is based on extremely large, complex, carbon-containing molecules. Life, in fact, exists because the carbon atom possesses exceptional bonding properties, and are able to connect with others in a variety of forms. Living systems are highly organized with specific spatial attributes and shape constraints. For something to be alive, it must be able to do the following:

- Have the ability to grow and reproduce (make more of itself)
- Inherit and pass on genetic intelligence
- Find and use food—metabolize (transform food to energy and raw materials)
- Discard waste
- Sense and respond to stimuli
- Adapt to the environment
- Maintain structural integrity and repair

Everything we need for physical survival is maintained by the life of our cells, either alone or with the help of other cells and nearby communities of cells. The cell engages in thousands of biochemical reactions each second to carry out the business of life. As we contemplate our tiny cells, we can see parallels in our lives: whatever they do to maintain life and survive is what we, too, must do to survive. We and our cells breathe, eat, assimilate and eliminate, recycle (we recycle things we use; cells recycle atoms), and regenerate energy (cells do so by recycling spent energy; at the level of the self, we do so by resting and replenishing). Though each of our cells is individual and independent, all cooperate as a community of one human body in a constant state of re-creation, balance, and communication. Cells need other cells to thrive; when isolated in a petri dish, a single cell cannot survive on its own—it will program its own death. Molecules and cells are continually removed and replaced,

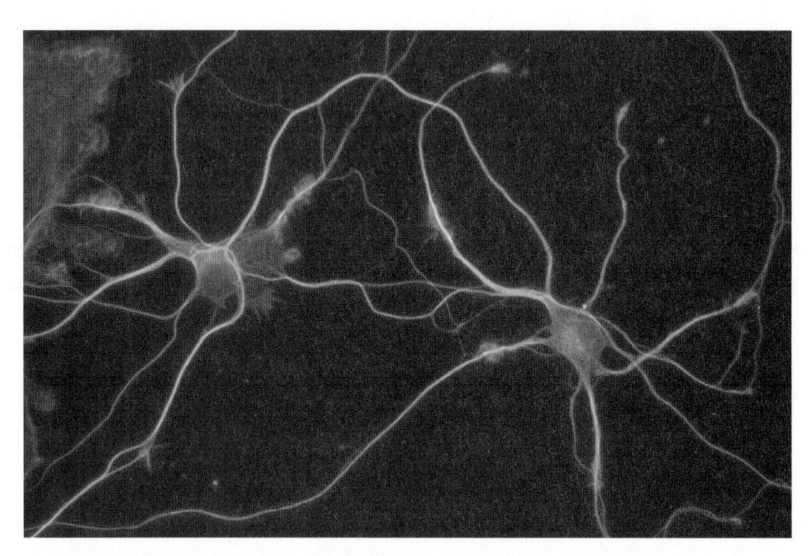

Figure 1.2 Two neurons; image by Dieter Brandner and Ginger Withers

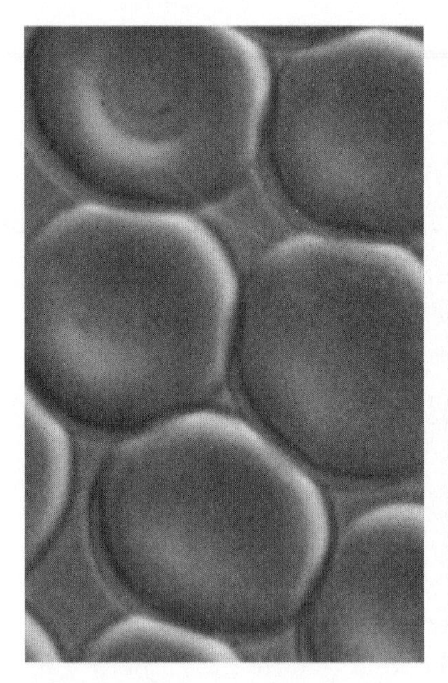

Figure 1.3 Human red blood cells

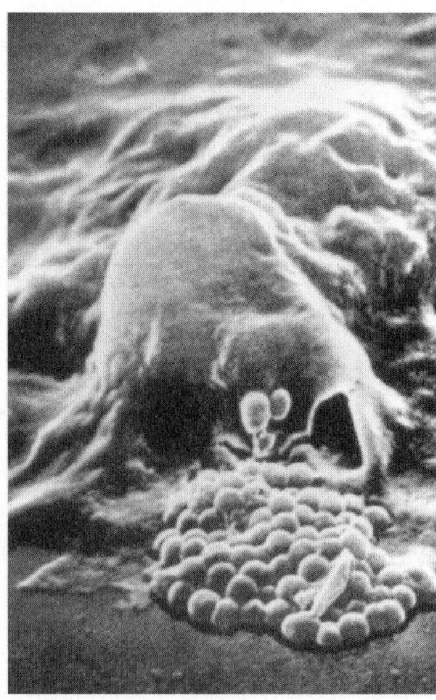

Figure 1.4 Scavenger white blood cells moving toward plastic beads

while the overall pattern and architecture of which they are a part are maintained—this is life.

Cells Are Us

The human body contains trillions of cells that originate from the intimate dance of sperm and egg into a single fertilized egg cell. During embryonic development from one cell to the many, each cell specializes and takes on unique features and responsibilities. Cells have different shapes and sizes, which influence and dictate their function. Cube-shaped skin cells stack together to make a covering for our bodies; it takes about a million skin cells to cover a single square millimeter. Discus-shaped red blood cells (see figure 1.3) speed through our blood vessels carrying oxygen where it's needed, while amoeboid-looking scavenger white blood cells (see figure 1.4) can squeeze through tissues seeking out dangerous invaders. One drop of normal human blood contains around three million red blood cells and five thousand white blood cells.

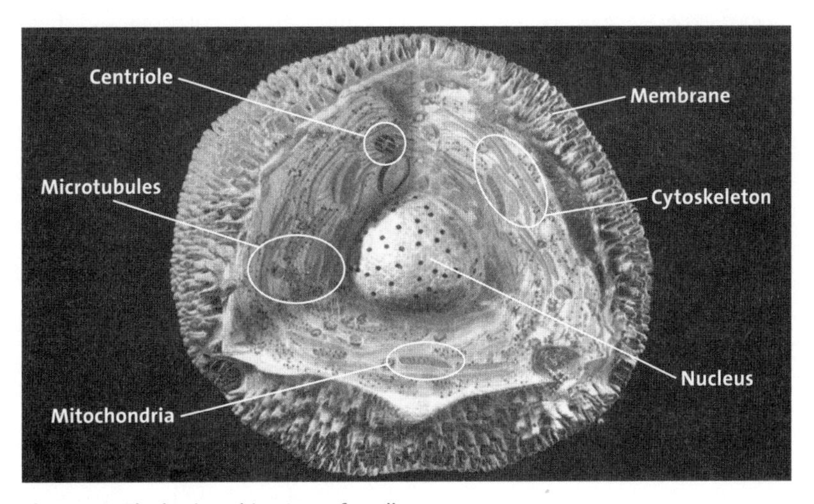

Figure 1.5 The basic architecture of a cell

Though cells differ in size, shape, and specialized tasks, they share some essential features and functions. A cell's basic yet revolutionary design, depicted in figure 1.5, enables the functions of physical life.

The outer surface of the cell membrane provides discriminating, resilient, and protective boundaries; the center core (nucleus) contains encoded genetic recipes; and the fabric of the cell (cytoskeleton) gives it a pliable structure, along with the ability to coordinate information, choice, and movement. Protein production encompasses the intricacies of the Golgi apparatus and endoplasmic reticulum. The spaceship-shaped energy generators are the mitochondria, and tiny grains called *lysosomes* are the dismantlers and recyclers of worn out and dangerous materials.

Consider the cell's survival skills as lessons from within. Every cell is in a constant state of maintaining integrity, balance, and fluidity. Our cells will carry us to the very end—how do we carry our cells?

Exploration

Cell Prayer: Your Cell Self

This is an easy exercise to discover your sense of boundaries and go further in your experience of "cell-ness." Read the following instructions and then allow yourself to engage in the suggestions offered.

Sit quietly, close your eyes, and breathe naturally. Now breathe into the place that feels like your center, reaching deep inside yourself with each inhale. Next, take your attention to your outermost boundaries, your outer edges. Where are they? Are they at the edge of your skin? Do they go beyond your physical edges? Notice the appearance of that boundary; is it solid, ragged, broken, or tattered? Now imagine that you are inside this cellular universe, a center core protected by whole, intact boundaries that are receptive to all that is good for you. If you like, you may even chant or hum out loud to connect all parts of your cell self. Voicing a sound with each exhale for a few minutes helps your cells resonate together. The sound "mmm" is particularly strong, and you will learn more about humming your cells in the next chapter.

Finally, take a moment or two to tune in and offer gratitude to this divine design that is you.

◇◇◇

Sanctuary: A Deeper Look

In its broadest definition, a sanctuary is a refuge, a safe haven. From a spiritual perspective there is an extra dimension of meaning: a sanctuary is a sacred place, a sanctified area, or a place containing an altar—a place permeated with spirit and imbued with the spark of life.

In Europe, Christian churches were often built on land considered holy, perhaps in a place where a miracle had occurred or a holy person was buried. The area around the altar came to be known as the sanctuary. In most modern synagogues, the main room for prayer bears the same name. The sanctuary is the innermost or holiest part of any place where people gather to pray.

Just like our cells, we humans build many kinds of "containers" for safe and sacred space. In the physical world, we seek sanctuary in wisdom circles, support groups, prayer circles, and knitting circles. We find sanctuary in our homes and sometimes even our cars. In indigenous traditions, healing circles include the whole community, and in the sacred tribal circle around Grandfather Fire, we come to pray, heal, and seek wisdom.

When we join in circles, we welcome love into our lives.

When we touch one another or hold hands,

we re-create molecular embrace and sanctuary on a grand scale.

Creating "Sanctuary" for Your Self

We can explore more secrets of the cell when we step into the sanctuary of our own life. Our innermost sanctuary is the heart of our cells. When we tune in to our sacred nature, we are able to recognize that we carry such a refuge within, and we can create an outer manifestation of this wherever we happen to be. For example, when I travel I bring a candle, incense, and a special cloth to make a hotel room "mine." At home, my garden is a sanctuary I go to on warm mornings to do my daily practice. Inside, I create altars to provide further reminders of sanctuary. There, a candle is always lit. No matter how small the space, we can always create reminders to help us remember the sacred. And in the space of our imaginations, we can always imagine that within each cell is an altar to life.

◇◇

EXPLORATION

Creating an Altar and Sanctuary

If you don't already have an altar or a place you think of as a sanctuary, take some reflective time to find such a place and consider what you would like to include there. Candles, stones, crystals, incense, plants, or pictures of your family, ancestors, or spiritual teachers are a few items you might want. A simple cloth and candle are more than enough to anchor sacred space and remind you to take time to enjoy and ignite the sacred from within.

Creating sanctuary is about embracing and honoring ourselves, taking the time to listen to what our inner wisdom has to tell us. It helps reconnect all parts of ourselves. When we take a step toward sanctuary and daily ritual, we help establish a new rhythm for ourselves and our cells.

◇◇

Secrets in Our Cells: Cellular Anthropology

The microscope opens deeper and bigger ways to understand life. It both uncovers the hidden workings of our cells and provides a doorway to unknown mysteries and other realties. In fact, in earlier centuries, microscopes and telescopes were considered to hold such power that most people were forbidden from looking through them, lest they alter their view of reality.

As a scientist, I learned that the microscope provides us with clues to our divine design and life teachings. As I noted in the preface to this book, I certainly wasn't looking for deep mythic meaning when I began studying cells and molecules. Nonetheless, they revealed a magical, mysterious realm that I learned to interpret through the eyes of a "cellular anthropologist."

Definition

Cellular anthropology? Yes. Anthropology is the study of human cultures and their origins; cellular anthropology looks at how the origins and forms of our cells may have contributed to human traditions, values, and art. A cellular anthropologist is a *code finder* who looks for clues to spiritual and social life hidden in our molecular structures as well as in the art and architecture we create. If you look again at the image of the medicine wheel (plate 4 in the center insert), can you imagine it also representing a cell?

Spiritual Architecture

I propose that our inner universe, including the cell, provides the template for spiritual architecture, a scaffolding upon which are built beliefs, values, art, and rituals. Humans, from the earliest times, created stories and symbols, music and prayers to move and touch their gods. People built edifices and altars, sang prayers, drew in the sand, danced, and discovered the basic forms and energy of the universe. Our cells contain secrets to such mysteries. What our cells do to hold and maintain life teaches us about divining life. In invisible sacred forms we discover the divine within and everywhere.

Reflection

Consider the sanctuary of your cells as a metaphor for your life as you contemplate the answers to the following questions:

Have you tended toward greater rigidity or flexibility as you have matured?

How discriminating are you with what you allow into your mind, heart, and body?

Where do you feel the most safe or nurtured?

Where or what is a sanctuary for you?

What do you hold as most dear and sacred?

If your soul had a container, what would it look like?

What does being self-contained mean to you?

Exploration

Attune to Your Cellular Sanctuary

Discover the cell's skills as operating instructions for your own life:

Learn to be in a state of self-creation.

Transform what needs to change.

Grow yourself.

Be fluid and flexible.

Aspire to the sacred within and around you.

Build an altar.

Find sacred space.

You have now begun to get acquainted with the workings of your cells; see how you perform the same actions in everyday living and understand the many levels of sanctuary that can be inspired by reflecting on the cell as a sacred vessel for life. As you continue through this book, each chapter will introduce a new lesson and another architectural feature of your cells. Take your time on this journey and allow your understanding to deepen as you travel. You are a sacred vessel.

Chapter 2

I AM–Recognize

It is an incredible journey from cell to self.

—CHRISTOPHER VAUGHAN *How Life Begins*

Now that we've become acquainted with our cellular sanctuary, in this chapter we explore how our cells "name" themselves. Here we meet many aspects of self: our cellular identification marks, the ways we may personally identify ourselves, how we identify others—even how we can participate in our own self-creation.

It is the job of our psyches to recognize ourselves and others as well as define our boundaries, while our cells do the same physically. Who would have thought our cells could hold such a crucial position? The family of immune cells takes on this essential task, and it is that family we will spend some time getting to know here.

Both cell and self say, "I AM," and when we fully recognize our cellular and soul connection to the sacred, we may say to ourselves, "I AM THAT I AM." Many religious traditions, starting with the Jewish people, have written that the name of God is "I AM." When we fully embrace who we are, when we say, "I AM," do we resonate with all that is? With all that is holy?

Who Am I? The Way In

When I first started teaching the mysteries of our cells, I asked my students to complete the following statements:

I am . . .

I want . . .

I have . . .

How would you complete the statements *I am . . . I want . . . I have . . .* ? All of these lead to questions people have posed for generations and that we, too, ask ourselves at various times during our lives: "Who am I?" and "Why am I here?" To begin exploring the subject of this chapter, recognizing self as distinct from others, take a few moments to reflect on the many ways you know and define yourself.

You have multiple "markers" of your identity. You have a name, a gender, distinguishing facial features, and a family tree. You have numbers that tag you: A birth date and a social security number link your identity to your "money self," to banks and other financial institutions and to your employment. These entities have probably assigned you at least one more number—you are employee B7834 or account holder 5483-14-070001. As technology has advanced, you have become further identified by an ever-growing list of numbers and passwords that allow you to gain admittance to the virtual world of the Internet as "you."

You have your profession or calling, your role in a family, your religious and spiritual inclinations and practices. And if they fit into your belief system, you have an astrological sun sign, a tarot symbol, and perhaps a lucky number. All of these are, or can be, parts of your identity, as are your beliefs and actions.

How *do* you recognize yourself? Is it by what you do in the world? By how people know and react to you? Is your sense of self externally or internally motivated? These are questions to contemplate as you begin to examine the parallels between your own sense of self and the self-identity of the cell. Knowing the self is like cultivating a garden: you have to be willing to explore the invisible and slow down to nurture your awareness. You need to be willing to work deeper than the surface. For your cells, however, the surface is the key to identity.

"Who am I?" is a question that can be answered by both our cells and our psyches, which together engage in an ongoing conversation to keep us safe. Body and mind share a common responsibility in self-identity, safeguarding us from danger and knowing what to trust. Both detect and protect our boundaries; the body's immune system, a scientific focus of this chapter on recognizing self and other, determines cellular boundaries and identities, while the nervous system navigates psychological ones.

Definition

Immune, from the Latin *immunis:* Exempt from public service or charge. Protected. Resistant to a particular infection or toxin owing to the presence of specific antibodies or sensitized white blood cells or relating to the creation of resistance to disease.

The basic job of our immune cells is to recognize "self" and "other" while collaborating with brain, gut, thoughts, beliefs, and hormones. The immune cells are sometimes referred to as a second sensory system, one that sniffs out danger. It is essential to add here that while our genes provide the information for crafting the physical and chemical aspects of our unique identity, our cells reveal those characteristics and our immune cells act on them.

Before we explore the specific activities of immune cells in some depth, let's take a closer look at cell identity (see figures 2.1 and 2.2).

How Our Cells Say, "I AM"

In the architectural design of our cells, the wonderful, fluid exterior membrane that we encountered in the last chapter reveals the cell's

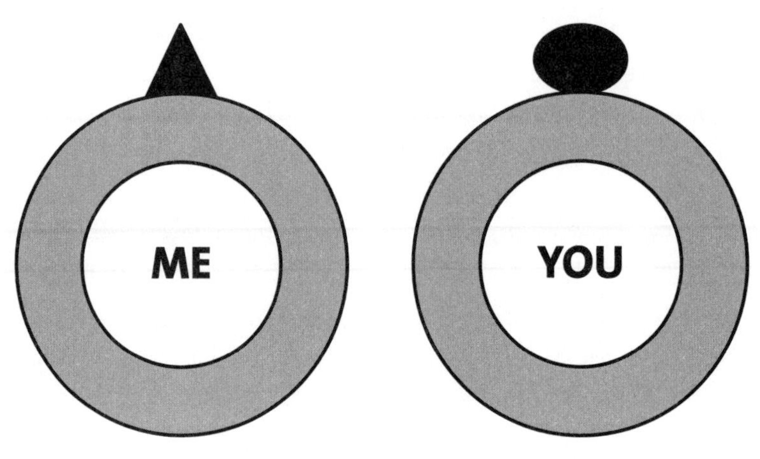

Figure 2.1 Cell as self, "me"

Figure 2.2 Cell as other, "not me"

identity. Just as you and I can tell a friend from a stranger by observing a person's external facial features, our cells do the same; each cell's "face," on its outer surface membrane, reveals uniquely identifiable features. Our cellular container is embedded with markings that enable cells to discern one from another. Blips and bumps on the surface are identification codes or passwords that mark "me" or self. These protein "signatures" on the cell membrane, akin to distinctive bar codes, reveal the cell's identity. These "me" markers also identify the cells as coming from you, a unique individual.

Like us, each cell has several layers of identity. In addition to "self" markings on the outer edges, cells carry "postmarks" or "zip codes" indicating where the cell originated and what it does: heart cells pump blood; white blood cells protect against intruders; red blood cells carry oxygen to all the tissues, and so on.

Recognizing Self and Other

When cells touch, their identification markings enable them to distinguish "self" from "other." A reading of "other," or *not self,* can signal either safety or danger: a threat that must be defended against. In the bigger picture of survival, "other" may be a threat if it is a pathogenic microorganism. Though it is well established that physical markings, shape, and touch are essential in enabling cells to recognize other cells and molecules, some scientists are now theorizing that molecular vibrations also play a part in recognition.

The way cells recognize differences is often explained with the analogy of a lock-and-key mechanism. The ID signatures on one cell are detected by decoding receptor sites on another cell's surface; the patterns or shapes of the two cells' markers fit together like a lock and key. The nature of the fit tells the cell whether what it has brushed up against is safe or not. Immunologists and biochemists have devised analytical methods to distinguish the many surface ID markings on cells, and this has proved highly useful in medicine.

The Medical Uses of Cell Identity

The original clinical use of cell ID markers was for red blood cell typing to ensure safe blood transfusions. Surface molecules on our red blood cells characterize them as either type A, B, AB, or O (see figure 2.3), and only like blood types can be safely shared from one body to another. (The exception to this is blood type O, called the universal donor type, since other blood types usually do not detect it as "other.")

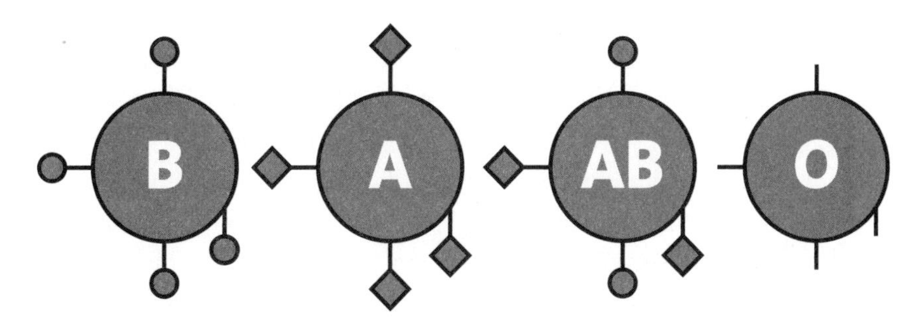

Figure 2.3 Red blood cell types

Later, modern medicine took advantage of a different set of identification markers, this time found on white blood cells, to make progress in organ transplantation. A person needing a kidney, for example, must receive one that has closely matched ID markers, determined by examining the white blood cells of both the donor and the person who will receive the organ. If the ID markings are very different, the donated kidney will be recognized as foreign and thus a threat (and sometimes cells of the donated organ will react against the new host); the immune system will attack the tissue and the transplant will be rejected. These ID markers are called human leukocyte histocompatibility antigens (HLAs).[1] There is an array of different antigen markers on a cell. Of the more than three thousand HLA markers, only six appear to be most important for organ, bone marrow, and stem cell transplants. All ID markers are considered to be antigens.

Definition

Antigen: Any substance that triggers a specific immune response. Antigens are usually protein molecules.

HLA Markers of Identity

Of course, these differences on our cell surfaces did not evolve in order to make blood transfusions or organ transplants possible. The biological role of the HLA codes, in contrast to their applied clinical use, is to provide the immune cells with self-protection. The different HLA markers reflect a range of immune responsiveness; some HLA markers indicate highly reactive cells, while others are slower to respond. Since all cells reveal markings of the "self," immune cells are trained early on—in utero—not to react against the self. Those that do are eliminated. Unfortunately, some self-destructive cells escape detection and later in life can lead to autoimmune diseases. Also, over a lifetime, the ability of cells to discriminate self from other may deteriorate. Now HLA testing is beginning to offer even more clinical benefits as it uncovers autoimmune tendencies and human responses to vaccines or drugs.

Definition

Autoimmune disease: A disease in which the immune system mistakenly recognizes part of a cell or a specific molecule as dangerous and attempts to eliminate the "not self."

Of the more than a hundred different autoimmune diseases, 75 percent occur in women. And they have become the third most predominant set of diseases in the United States after heart disease and cancer.[2] Examples of common autoimmune diseases are type 1 diabetes, rheumatoid arthritis, Graves' disease, and lupus.

Reflection

When have I lost the ability to discriminate between people, places, or behaviors that are well matched to me and those that are not?

Failures of Self-Recognition

In autoimmune diseases, the recognition of "self" is compromised: our own cell or protein molecule is no longer identified as ours—it has become the enemy. Certain HLA patterns signify the likelihood of developing an autoimmune disease and indicate hyperactive immune cells.

In addition to mistakenly recognizing the self as other, in some autoimmune disorders, this inappropriate response fails to be suppressed; thus, autoimmune diseases can represent an error in both recognition and regulation. Across the many different autoimmune diseases, mechanisms of failed recognition vary. Some tissues or proteins lose or change their ID markings and are now seen as "not self"—their identity has been hijacked. In other situations, it's a case of mistaken identity. The immune cells make an error and misinterpret self markings as if they are those of a foreign intruder.[3]

When immune cells recognize any cell or substance as a dangerous "other," they attack and destroy it. An example of this occurs in multiple sclerosis (MS), where a protein of the myelin sheath that protects the spinal cord is mistakenly recognized as foreign and attacked; over decades, the nervous system will be slowly destroyed.[4]

While it seems that autoimmune diseases are reaching epidemic proportions, this may be in part the result of our increased ability to recognize autoimmunity. Thus, in medicine, cell identity is a vital area of study. In time, scientists may be able to help cells regain their correct identity or eliminate those that have lost their ability to know the difference; then we will be able to reverse the current trend of increased autoimmune disease.

Parallels in Consciousness

Self-recognition by the cell has parallels in the psyche. Dr. Jeanne Achterberg, a pioneer in psychoneuroimmunology and imagery, has encouraged individuals with autoimmune diseases to ask themselves such questions as "How have I lost myself?" and "How do I need to recognize myself?" The implication is not that the person's sense of self or thinking patterns *created* the disease; rather, considering these questions can add another dimension to healing. We don't need a diagnosis of an autoimmune illness to feel that, at some time in our lives or perhaps multiple times, we've failed to recognize our true selves.

The Scent of Self

We inherit the genetic programs for our cellular HLA identity markers from our parents. These markers not only indicate the quality of immune responsiveness, they also provide an unexpected distinguishing characteristic reaching beyond the invisible microscopic "self."

Picture a bloodhound looking for a lost child based on the scent lurking on the child's blanket. The dog is actually detecting "cell droppings" that make up the child's unique smell. Our cells' HLA ID markers are shed into our sweat, urine, and saliva, producing a unique aroma. So powerful is this identifier that within the first twenty-four hours of life, newborn babies can distinguish their mothers from other women based on smell.[5]

We adults can smell differences too, subtle though they may be. Fascinating research originating in Switzerland demonstrated that our sexual attraction to another may be traced in part to the aromatic trails our cells exude. First, scientists observed that mice that looked identical only mated with mice of a different genetic strain. How could one mouse know another mouse's genetics, the scientists wondered—by the other mouse's smell? Extensive research using an "electronic nose" to discriminate the volatile components in mouse urine indicated that urine from different genetic strains showed different odor types, and this was reflected in mouse mating behavior.

Based on these animal studies, in 1995 Swiss biologist Claus Wedekind explored the theory that HLA differences were connected to sexual attraction between men and women. Male college students were given new white T-shirts to wear for two days. During this time, they were not allowed to use deodorant, aftershave, or scented soaps. Afterward, the shirts were placed into small cardboard boxes, and female students were then asked to smell each shirt and rate it: pleasant or unpleasant, from a "sexy guy" or someone who was not sexually appealing. They did not see the guy. The results showed that when a woman indicated the smell was pleasant, her HLA markers differed significantly from those of the man whose shirt she liked, while the shirts she deemed unpleasant smelling had been worn by men with similar HLA markers. Good smell equaled different markers.[6]

This makes sense from an evolutionary standpoint. Remember that HLA ID markers indicate a range of immune abilities and immune responsiveness. If we mate with someone who has very different self markers from ours, our offspring will inherit a more diverse immune capability than if we pair up with someone having similar markers. Interestingly, the researchers found that a woman's ability to discern these differences is lost when she takes birth control pills.[7]

A number of years ago, when I was teaching this correlation between attraction and smell, I heard remarkable personal stories that further supported the conclusions from this research. One woman decided to get pregnant to see if a child would help her troubled marriage. Once off her birth control pills, which she had taken since before getting married, she found her husband's smell offensive. So strong was her revulsion that she decided to end the marriage—and, of course, she believed something was wrong with her to react in this way. Hearing the underlying biological story—that her birth control pills had blunted her perception of the couple's cellular, "smellular" differences—helped her understand that final trigger for divorce. While there had been plenty of conflict in the marriage already, the story her nose told her supplied the deciding factor.

Another woman told a story that could be a fertile area for research. She had adopted a baby girl about eight months earlier and felt that the baby had not bonded to her. "Could it be because of smell?" she asked. It's a good question. Is it possible that we could help adoptive moms and babies connect more closely if the adopting mother had a piece of cloth containing the birth mom's perspiration?

Smell, Self, and Memory

Fundamental to self-identity, both immunity and olfaction function to detect molecules that belong to us and those that don't.[8] And both systems are also linked to memory. Certain kinds of immune cells can actually remember a previous invader and protect us should it attack again. Through certain smells we remember loved ones, traumatic events, and other experiences that are important to us. My mom, for example, wore Estée Lauder perfume whenever she got dressed up. Smelling or remembering that scent always brings my loving mom—part of my memory and identity—back to me.

Another example of the connection between smell and memory occurs in Alzheimer's disease; loss of the sense of smell often precedes memory loss and the loss of the sense of self.[9] Yet loss of smell does not always signal Alzheimer's disease. A zinc deficiency also can result in a loss of smell and taste as well as diminished immune function.[10] Have you ever taken a zinc lozenge to help heal a sore throat? Zinc is a potent immune activator and may restore a lost sense of smell.

Physical Markers of Self (Biometrics)

In addition to cellular markers, our bodies provide other self-identifying characteristics, including fingerprints, retinal prints, voiceprints, and genetic footprints. All are unique; most do not change with age or health status. (On rare occasions with certain neurological syndromes, fingerprints may change, while retinal patterns can change in people with diabetes.)

An interesting phenomenon occurs in a very small population of women with chronic fatigue syndrome (CFS): they lose their fingerprints![11] Chronic fatigue syndrome is a complicated body-mind illness involving a debilitating fatigue that lasts longer than three months, an exhaustion so profound that people are unable to do the things they used to do. And in this respect, they have indeed lost part of themselves—the people they have been in the world. CFS is often diagnosed as "all in your head," and there is no clear universal treatment, but we know it represents failure of immune balance, for frequently the individual's blood reveals a hyperactive immune state. Often a viral infection precedes the onset of the disease, and the immune cells continue to react as if the virus is still present long after it has passed.

The Cellular Sleuths Detailed: The Immune System, Protector of Identity

The protector of cellular identity, the immune system, is one of the most complex networks in the body. Here I offer a brief yet fairly technical overview of the science of how immune cells work. It is truly a miraculous process that continually affirms my sense of the divine nature of the cell.

Our vastly diverse immune network is a combination of interactive cellular communities, magic potions of molecules, and the individual's daily experience: attitudes, emotions, stresses, and happiness.[12] The immune organization is truly a body-mind experience. The primary cells involved are the white blood cells, and the main molecules are called antibodies and cytokines. Add in the workings of the brain, hormones, and lifestyle, and you still have only *part* of the immune network.

The white blood cells, designated as the primary sleuths, sense danger lurking in the microenvironment. There's actually a whole family of white blood cells involved in this mission: neutrophils, eosinophils, lymphocytes, dendritic cells, and monocytes-macrophages. In the grand scheme of cellular self-protection, once the immune cells recognize a foreign invader as "not self," they will attack and destroy it. (It is worth mentioning that in more than thirty years of teaching about the immune system, I have repeatedly asked students to come up with a metaphor for this behavior that does not involve warfare. No one has yet—perhaps you will.)

Defensive Posture

There are basically two groups in this family of immune cells: the innate primitive phagocytic scavenger cells and the more sophisticated, educable lymphocytes. Our innate cells (neutrophils and monocytes-macrophages) are responsible for nonspecific, quick-action immune responses. On the other hand, lymphocytes are responsible for the long-lasting, learned immune responses. The more primitive scavengers, or phagocytes, are the first to detect foreign agents such as the larger microorganisms. These cells "eat" other threatening cells or cellular debris, hence the name phagocyte. They slither around, changing shape to get into the smallest spaces and tissues, detecting any invader that has breached the sanctuary. They can detect bacteria, fungi, and even dust particles in the lung. Phagocytic cells defend us against infection yet never remember what they've done.

The second arm of immune defense, the lymphocytes, will come in to complete the job as well as eliminate the more hidden, smaller viruses, and they will remember these intruders' identities. This more sophisticated acquired immune response depends on lymphocytes, which are educated throughout our lifetime. You will read more about their role further on.

All of these cells are endowed with surface markings that enable them to recognize and respond to "not self."

An Inflaming Scenario

You're in the garden trimming a rose bush when a hidden thorn snags your finger. Ouch! A drop or two of blood oozes from the wound, and millions of cells are called to rescue and repair the tiny tear in your skin. The fluid plasma in which the blood cells swim floods the area, washing out and diluting any toxins. The white cells hear the alarm. Neutrophils pick up signals from invading microorganisms, recognize them as "not self," latch onto them, and gobble them up, killing and digesting them in the process. To nourish these immune cells, red blood cells rush in to bring oxygen and remove waste. Platelets, another kind of tiny blood cell, help wall off the area so that the invader and any accompanying toxins stay put. Within the first twenty-four hours of the thorn's "attack," the other scavenger population, the monocytes, come in to finish the job.

Bacteria and fungi are the pathogens most readily eliminated through this process. Once they are gone and the area is cleaned up, chemical signals instruct the cells to build new tissue.

This entire scenario is one you have experienced again and again: it's called *inflammation.* Next time you get a cut, insect bite, bruise, or burn, watch what happens. The area reddens, swells, heats up, and becomes painful; these are the four classical signs of the inflammatory process—signs that your immune cells are working.

Inflammation is the most basic immune response. Whether a foreign organism or an irritating dust particle triggers the assault, white blood cells are aroused to handle the situation. Anything that breaches our physical boundaries can damage the self and excite these cells into defensive action.

Unfortunately, this response can get out of balance; scientists are discovering that many diseases are inflammatory in nature.[13] For instance, though we have become convinced that high-fat diets lead to coronary artery disease, compelling research indicates that inflammation of the blood vessels is a major underlying factor.[14] The tissue-damaging effects of diabetes are also partially attributed to inflammation, as are the long-range dangers of obesity.

To mitigate inflammatory assaults, be physically active and enjoy a diet full of fruits and vegetables: rich sources of molecules called antioxidants that can help lessen the dangers of excessive inflammation. When you learn about and understand cellular behavior, you also learn what you can do to help lessen or prevent the damage of an out-of-control cellular response. Your cells will love you for it!

The Second Call to Arms

The other group of white cells, the lymphocytes, protects us against viruses and continued assaults by other microscopic invaders. The lymphocytes known as T, B, and NK cells provide different collaborative roles in protecting us. Thymus-derived T cells are the regulators:

T helper cells activate the immune response; T suppressors turn it off. The B cell family produces antibody molecules to coat and neutralize antigens—antibody-coated antigens are easier to eliminate. And for the most part, B cells require the support of T cells to carry this out. "Natural killer" NK cells are the earliest defenders against viral infections, killing virus-infected cells directly within hours of attack, even before the rest of the immune collaboration takes place. They also appear to be sensitive indicators of psychological and lifestyle influences on immune health.[15] This means their activity is easily measured in the test tube. Feeling lonely? Your NK cells may become sluggish. Meditate or watch a funny movie and you pep up these cells.

While the scavengers are born with skills to immediately recognize and remove microscopic threats, the T and B lymphocytes need time to develop their selective powers before becoming fully armed to eliminate the invader (called acquired, learned, or specific immunity). Another distinction between lymphocytes and scavenger cells is that most lymphocytes can develop memory. As I mentioned earlier, the scavengers (monocytes and neutrophils) simply do their job, remembering nothing afterward. In contrast, memory lymphocytes, made during an infection or immunization, protect against future assaults by the same offending organism. That's why people who receive the measles vaccine are protected for a lifetime; a population of their lymphocytes can instantly recognize and eradicate any measles virus that might assault them. One of the things that continually amazes me about the immune network is the number of fail-safes and alternate means to eliminate danger that are built into the system. And the cells do not operate in a vacuum: what we do can help or hinder their abilities. We live in a truly collaborative relationship with these cells, a partnership dedicated to keeping us healthy.

The "immune dance" of the acquired or learned immune response is an astonishingly complex choreography that brings together all aspects of body, mind, and molecules. It shows that our cellular universe is a very cooperative and collaborative one protecting our survival. Our lifestyle, nutrition, and physical activity as well as stress all play a part in this dance; long-term chronic stress can slow down immune responsiveness, while relaxation strategies can improve it.[16]

Now let's walk through the steps of this dance.

The Eleven Steps of the Learned Immune Response

1. *Recognition of "not self"—an invader.* The scavenger monocytes and dendritic cells begin the dance. Once cells detect the invader, an almost magical performance is choreographed.
2. *Dismantling the invader.* Once recognized as a danger, the offending organism is broken into pieces that stimulate immune reactivity. These stimulating pieces are called antigens. The dismantling cells, called antigen-processing cells (APC), include monocytes-macrophages and dendritic cells.
3. *Recognition markers point out the invader.* The antigen is moved onto the surface of the antigen-processing cell, like a flag. The cell circulates, looking for a T helper cell that recognizes this offending antigen. T helper cells are endowed with the ability to recognize about a million different antigens, though each specific T cell recognizes only one.

4, 5. *Molecular messaging—a feverish response.* Invasion! Once the appropriate T helper cell is found, it receives a chemical signal from the antigen-processing cell to make more copies of itself. These chemical signals are called cytokines and interleukins, which are molecular messages between cells. IL1 (interleukin 1) signals the T helper cell to make more helpers (5) able to recognize the specific antigen. The T helpers choreograph the rest of the dance. In addition to affecting immune cells, IL1 also travels up to the brain, increasing body temperature and making you sleepy, helping you conserve energy to fight the infection. For many invading organisms, an elevated body temperature is lethal. In fact, fever is another hallmark of a working immune network. Though we often interpret a fever as "something's wrong," on the contrary, it is telling us that our immune network is working fine. Studies of people with influenza virus infections who took aspirin to lower their temperature indicate they actually experienced longer-lasting symptoms than did people who took no aspirin. In fact, aspirin lowers a fever by lessening the immune response.

6, 7. *T cells look for B cells.* The T helper cell, now carrying the antigen flag of the invader, goes on the prowl to find a B lymphocyte that recognizes the same antigen. The B cell spots it and sends another molecular signal (7) so that more B cells are produced that recognize this specific danger.

8, 9. *B cell expansion and antibody production.* As this population of B cells expands, they mature into plasma cells, which manufacture antigen-neutralizing proteins called immunoglobulins or antibodies (9). There are several classes of immunoglobulins: IgG, the most predominant form, is present in the blood, while IgA is primarily in the saliva and gut. IgM is the first form produced, whereas IgE is made in response to parasites and allergens.

10. *Suppressor cells halt action.* Once the immune defenders have duly eradicated the invader, another set of molecular signals trigger the all-clear switch; T suppressor cells halt the immune response.

11. *An important handshake.* You may wonder: if immune cells carry the marker of an invader, why aren't they recognized as "not self" and attacked? Self cells "hold hands," recognizing each other; hence, no attack.

The immune network is preeminent for its collaboration between the physical cellular universe and our more etheric consciousness. It serves to define the physical cellular self while being influenced by our invisible beliefs, attitudes, and psychological perceptions of self.

Who Am I? Discovering Self

Now that we have witnessed the uncanny intelligence of our cells in identifying and protecting self, let's shift our conversation and think about what gives us as whole, multidimensional, and conscious human beings our positive sense of self, and how we may strengthen it.

A core belief I've held since I was a child is that every one of us has a special purpose for being here. How are we meant to use the spark of energy given to us at the beginning of life? To love? To work? To nurture the land or a child? To dream and complete a creative project? Each of us likely carries a seed of inspiration about who we are and why we are here. But what if we don't have or hear that inspiration—then what?

There was a time when my inspiration and sense of mission had disappeared. "I" was truly lost and chose to travel a new path to uncover my spiritual self, apprenticing with a shaman. We worked with sound and nonlinear technologies to access parts of myself I had been unaware of. With practice, commitment, and community, I recovered more of me, and it was this path that offered me the greatest integration of self. I raised the old questions: Why am I here? What is my unique gift that would make the world a better place and benefit others? It is what most of us ask at some time in our lives, and we often need nontraditional means to access that information. I would now call them sacred technologies.

In my apprenticeship, I "stalked power" in order to be able to fully know and express who I am. In shamanic or indigenous traditions, power is called "medicine" and each person's medicine represents his or her individual gifts, purpose, and power for healing the tribe. I found that one of my gifts was sound.

When we uncover our power and begin to put it into action, we truly are ourselves. It is then that we make the greatest contributions and use our gift of energy more wisely. We have so many pathways to tap into, know, and integrate the self. Since sound is my given medicine, here I offer ways you can use it to enliven your sacred cell self.

Engaging the Sense of Self: The Power of Sound

Arden Mahlberg, a psychotherapist in Madison, Wisconsin, uses sound to help his patients rediscover their core sense of self. He tells the phenomenal story in *Music and Miracles* of his first clinical experience with the sound "mmm." One of his patients, a thirty-five-year-old man, felt inadequate and trapped in a safe job that he hated. He was depressed, and he feared failure. Mahlberg prescribed sound: imagining the "mmm" sound and then humming it. At first, the image and sound provided the man with more restful sleep—who couldn't benefit from that? Then, after several months' practice, he found he had created a radical shift in his identity. He gained a new sense of security and was able to change jobs without worrying about the outcome.[17]

Now Mahlberg often uses this sound strategy with patients who've made little progress in "talk therapy." Humming helps them find themselves. He asks patients who feel lost to first imagine the sound "mmm" and then hum it for between five and twenty minutes every day. In his experience, over time, these individuals become more assertive and confident, making clearer choices and decisions. Frequently, as they view themselves (the self) differently, their lives take on new rhythm and energy. In essence, they rediscover themselves and their true nature. Do you ever feel you don't know who or where *you* are? Here is an exercise for you.

Exploration

The Sound of the Self

Explore this experiment to strengthen your sense of self. Even if you feel just fine as you are, this can be an interesting enhancement to your well-being.

Set aside about five to ten minutes in a place where you feel safe and will be undisturbed. Sit in a relaxed position. Close your eyes and imagine or picture the sound "mmm." You can simply imagine the letter *m* or pretend that you are hearing the sound. Another option is to make the sound silently. Do this part for a few minutes.

Now you can move on to the next step and practice aloud with sound. Take a deep breath, inhaling through your nose. Your mouth is closed.

Exhale, letting the breath out through your nose while humming "mmm." Allow the humming "mmm" to last as long as your outbreath will permit. Now take another deep breath in, and once more sound "mmm." Feel the sound vibrate your bones and tissues. Where do you feel the sound first? Repeat at least nine times to begin to feel the benefits.

Experiment with using high and low tones, loud and soft. Do you feel the sound in different places when your tone is low and when it is high? Feel the difference when you have your teeth touching as you hum.

Build your practice over the next few days until you are sounding "mmm" for at least five minutes. Be silent for a while after each session. Notice how you feel. You will find that five to twenty minutes of this changes your day. Keep a journal of your experience. Do you notice any physical changes, any new awareness or sensitivities? What about your sense of self, your boundaries?

The great teachers tell us that it takes at least twenty-one days to change a pattern. My suggestion is to begin the humming practice three days in a row, working up to twenty-one.

Warning! Do not do this while driving a car. People often enter a deeply altered state as the harmonic hum slows down both brain waves and response times. This is excellent preparation for imagery or meditation, not for activity.

How Can Humming Possibly Help Recover the Self?

Our human knowledge of the therapeutic effects of sounding "mmm" goes back thousands of years. Socrates said that listening to humming kept him content, focused, and secure in himself. Kabir, the Sufi poet, said the sound moved him to ecstasy. The Hupa Indian tradition teaches that when we wake up feeling out of sorts, humming will help our spirit return to us.

Mahlberg chose the sound "mmm" because he postulated that it is the archetypal sound of the self—there is plenty of evidence for this. Regardless of our cultural roots, as infants, we make the "m-m-m" sound while sucking. When sounded with an empty mouth, "m-m" becomes "mama," globally the most common first spoken word. In many languages, the word for mother emphasizes the "mmm" sound. Parents naturally hum their children to sleep. We purr as we take in pleasing nourishment: "Mmm, mmm, good." Is it a coincidence that "mmm" is a sacred sound in many traditions: amen, aum (om), shema, shalom, salaam?

In the Hindu Upanishads, the "mmm" sound is said to provide a fundamental distinction between me and not me, self and other, inner and outer. In other sacred traditions this sound provides a link between heaven and earth. Aum and amen are sounded at the end of a prayer or spiritual verse. Shaman and anthropologist Angeles Arrien describes the sound of the letter *m* as the consonant sound that grounds our prayers; while "ah" opens us to heart and spirit, "mmm" brings us back to earth.[18]

When we hear and make a repetitive sound, our brain waves slow and become similar to the delta brain waves of restorative sleep, which are also the first detectable brain waves in the fetus. When we hum, are we re-creating the peacefulness of womb consciousness—making a new safe place?

Humming vibrates the body, from inside the cells through flesh and bone, the strings of our cells, and into the mind. It attunes us, helping most of our cells to resonate together. This affords us the perfect time to align our ideas of self and intent.

If you already engage in a body practice such as yoga, tai chi, or qigong, try adding the "mmm" sound to your practice. You may find that it helps you further embody yourself.

Sound at Work

Many people find that repeating a sound as a single, monotonous tone induces and deepens the meditative state. In fact, I have worked with clients who were unable to meditate until they sounded "mmm" or did some kind of chant.

Witnessing such emotionally centering effects, I became inspired to try sound with people who had immune illnesses that featured failures in cellular identity. For a period of a year, I worked first with a group of eight women suffering chronic fatigue syndrome, a baffling disease mentioned earlier in this chapter. This is a condition with an ill-defined set of symptoms that result in diminished cognitive abilities, memory, physical strength, zest for life, and immunological integrity. Frequently, the people struck with it are high-powered overachievers, often women in roles that have traditionally been considered male. The syndrome certainly represents "losing the self" at many levels: you can't do what you used to be able to do; you don't remember what you've read; your energy is exhausted. You sleep but wake unrefreshed.

The women in my group practiced the "mmm" sound to see if they could recover more of who they once were. Every one of them reported that making the sound for even a few minutes gave them more energy and that over time they felt better about themselves.

As I began integrating sound and "mmm" into other groups, I repeatedly discovered that it was a powerful tool for enhancing well-being. One man with an autoimmune illness, for example, had unrelenting muscle and joint pain because his immune system had been progressively destroying his cells. The only strategy that relieved his pain was sound: he used humming with the "mmm" sound as described above as well as other tones.

Every cell, molecule, and atom vibrates, moves, and hums. Radio astronomers detect a constant humming sound everywhere in the universe. It is said that if the sound of our atoms is heard, it resembles OM (aum), the sacred Hindu sound. Yet aum is more than an ordinary sound in the Hindu tradition; it represents the name and manifestation of God, or divine consciousness. Joseph Campbell described the inner humming sound as the primal energy and vibration of which the universe itself is a manifestation. And if sound manifests or organizes the universe, doesn't it follow that humming can help us manifest our individual universes?

Reflection

How do I recognize myself?

How am I recognized?

How do I fail to recognize myself?

How do I recognize others?

What aspect of my self needs strengthening?

Who Am I?

To give you examples of all the ways we can recognize the self, here are some of the ways I answer the question "Who am I?" I am atoms, molecules, cells, thoughts, feelings, ideas, Sondra, daughter, mother, grandmother, friend, lover, work, spirit, love, sound.

In terms of metaphysical and symbolic traditions, from astrology I am a Libra-Scorpio; from tarot and numerology, the priestess, justice, an 11 and a 2; and from another metaphysical system, I am the queen of hearts. Consider: If one ascribes to metaphysical qualities, how do they add to our identity and life purpose? Where do such mythic or esoteric labels support, enhance, and expand our self-identity?

Body Prayer

Anchoring—A Declaration of Self-Creation

The following exercise can help you gain a solid sense of self and serve as a tool for intentionally creating and re-creating the self you want to be.

Choose a place that feels like a sanctuary for you. Take a moment to "get there." Then say a few words, silently or aloud, to thank your cells and the great mystery for all that you are, have, and know.

Feel your feet on the earth, grounded and anchored. You can gain strength from the earth's energy when you feel your feet upon her. You can also imagine that your head is connected by an invisible cord to heaven. Rock a bit back and forth on your feet until you feel solid on the ground.

Imagine you are a rope held between the earth and the heavens. Make circles with your waist, rotating belly and hips, moving slowly as you would with a hula hoop. Your shoulders remain parallel to the earth while the middle of you leads the motion. Move like a rope being held at both ends. Continue until you feel anchored. Then change direction. You may discover that circling in one direction feels easier and more natural than the other. Which way is more comfortable for you? Now add the humming sound "mmm" as you root and spiral, connecting to the earth and your cells. Stay with this for a few minutes.

Now, firmly rooted on the earth with your feet anchored, raise your arms to the heavens, flexing your hands at your wrists. Touch your hands to your heart and belly and then bend down to touch the earth. Do this at least three times.

Make this a prayer of intent. When you raise your arms, ask for guidance or what you need to know to take your next step in life. Or, as you reach up, give gratitude for all that you are and have. When you touch your belly and heart, you commit to your intention or your higher self. As you reach to the earth, you plant your prayers of intent, perhaps saying aloud that you'll do whatever it takes to make this seed grow.

When you become rooted and enact the three movements, you are in essence declaring an act of self-creation.

Toward an Integrated Self

We began this chapter by examining an almost universal religious statement: I AM THAT I AM. Now let's revisit that idea as we might find it in our daily lives.

Our cells, when they are healthy, know exactly who they are, while our psyche may consider the question in an ongoing fashion: "Who am I?" "Who am I now?" As we journey toward maturity and transformation, accepting who and how we are in the moment takes us to a deeper appreciation of I AM THAT I AM. *Acceptance is part of our human recognition of self.*

Recently a student called me to share her revelation about her cells and self. Her walk one morning was particularly difficult. It was literally an uphill battle for her and her cells. She told me that on the steep trek her cells were screaming, feeling stretched to their limit; they hurt and wanted to stop. Then her consciousness (what I've been calling mind and psyche) revealed its role in the walk: we all benefit if we go past the struggle to reach the top of the hill. That's when she understood the collaboration between cells and consciousness, matter and mind. The cells can do their thing unaided by mind, yet when the mind accepts how it is and says let's go a little farther—let's not quit yet—the entire being benefits, cells included. The struggle adds to a new sense of self.

When we accept and acknowledge that our choices influence our cellular health, we can embrace our cellular nature as a sacred chalice or vessel we have been given to take care of, to nurture. We know that we are one.

In the next chapter you will discover how our consciousness sends messages to each cell and how cells engage in conversations with one another.

Chapter 3

Receptivity–Listen

Our smooth-muscle cells . . . work away on their own schedules . . . opening and closing tubules according to the requirements of the entire system. Cells communicate with each other by simply touching; all this goes on continually, without ever a personal word from us. The arrangement is that of an ecosystem.

—LEWIS THOMAS *The Lives of a Cell*

As we continue to explore the divine nature of our cells and ourselves, in this chapter we move from "me" to "we." Our cells have much to teach us about perceiving one another, listening, being heard, and the value of connection. And how our cells "listen" is through receptor sites or "antennae" on their outer surface.

Cells Talk

If we were able to listen to our cells' conversations, do you think they would be talking about how to wrangle a raise at work or who did better on an exam? No! We'd hear them discuss moving faster or slowing down, breathing a little more deeply, pushing forth and pulling back. We'd hear them ask, "Which message shall I respond to?"—there are so many.

The better we understand how our cells manage our ever-changing internal, alchemical environment, the better will be our relationships with ourselves and each other. And the greater appreciation we will gain for what magnificent creatures we are.

Again, as in the two preceding chapters, we begin with the surface of the cell, the membrane (see figure 3.1).

We have learned that the cell membrane carries distinguishing marks revealing its own nature—I AM. The membrane also holds the ability to communicate with other cells. Communication depends first on the membrane's antennae or receiving sites (receptors), which have the uncanny ability to recognize specific incoming molecules. If the molecules are a good fit, our attentive cells will embrace them. This is similar to the design requirement for the immune collaboration between antigens and membrane receptor sites described in chapter 2; cells use the same mechanism to recognize molecular messages from a changing environment. Nobel laureate Christian de Duve has named this phenomenon *molecular complementarity.*[1]

Figure 3.1 Diagram of cell with self markers (triangles) and different receptors (the other shapes)

Biological information transfer is based on chemical complementarity, the relationship that exists between two molecular structures that fit one another closely—a dynamic phenomenon as the two partners are not rigid. When they embrace, they mold themselves to each other to some extent. Embrace leads to binding.

—CHRISTIAN DE DUVE *Vital Dust*

What messages must our cells attend to? Chemical signals of danger, a need for new resources or energy, speed up, slow down—there are molecular messages to rest and relax, renew resources, and recycle. Because of instantaneous recognition, our cells can respond in nanoseconds to keep us protected, moving, and replenishing. But how?

The Receivers

Each cell's surface is studded with thousands of receptor sites whose function is to detect information essential to its survival. Any given receptor is shaped so that it can only accept certain messages. The same lock-and-key analogy we have seen before applies here as well; only if a cell has a receptor able to recognize a particular chemical signal can it respond. This boils down to both the receptor and the initiating message having the correct spatial, geometric, three-dimensional structure relative to one another. Some scientists say that the vibrations of each—receptor and message—also play a part in this dynamic conversation. The receptor is not fixed; it molds itself around the message, holding on.

Recall that the cell membrane is made up of a double layer of fats. These are called phospholipids, and they merge to form what we can think of as a stable "soap bubble." The protein receptors float in this lipid surface layer while some of them straddle the membrane as many as seven times, touching both outside and inside the cell. The receptor grabs its matching molecule from outside the cell with its "hands" and then moves its "feet" on the inside to tell the rest of the cell that a message has been received. The cell can respond to the message, provided that all the other necessary mechanisms *inside* the cell are engaged and turned on. This involves a *conformational* change in the physical structure of the receptor—that is, the receptor *changes shape* when it contacts a molecule it recognizes. In *The Biology of Belief*, Bruce Lipton writes that this membrane activity constitutes the brain of the cell. Yet cellular intelligence requires much more than membrane receptors, something you will learn more about in the next chapter. The ultimate aim of a molecule binding to its receptor (if molecules have goals) is to turn the cell on for a specific activity. Examples of cellular activities include making more energy, slowing down, and contracting. You can envision the interaction as a molecular embrace: the receptor has to "hug" the signaling molecule. If the squeeze is too tight, the "on" switch may stay on too long; if it's too loose, the cell might not even know it's supposed to do anything. It's the "just right" connection that encourages the cell to change what it's doing and take a different action.

One of our first insights into the importance of cell receptors began around sixty years ago when researchers examined how adrenaline works. Adrenaline, also known as epinephrine, is a molecular signal of distress that tells our bodies to get ready for action. It ensures that there is enough sugar in the blood to keep things going by facilitating liver cells to free glucose from its storage form, called glycogen, and release it into the blood. Stress (the fight-or-flight response), which can be triggered by an acute physical or mental challenge, creates a chemical drumroll that moves energy to our muscles, increases blood pressure to get oxygen to the cells, and accelerates everything needed for immediate survival. If the cell receives a message of danger, it enables us to fight or run away. Almost every cell type in the body has receptors for adrenaline, though each responds in its own specific way. Heart cells beat faster in the presence of adrenaline, while cells in the pancreas stop secreting insulin. Every part of us has a job to do to deliver us from danger.[2] To give you a sense of different receptor shapes, compare the form of adrenaline (see plate 8 in the color insert) to the form of caffeine (see plate 9 in the color insert). You can easily see that cells need different kinds of receptors to respond to the diversity of chemicals.

If you were to assign a different color to each of the receptors that scientists have identified, the average cell surface would appear as a multicolored mosaic of at least 70 different hues—50,000 of one type of receptor, 10,000 of another, 199,000 of a third and so forth.

—CANDACE PERT *Molecules of Emotion*

Masquerading Molecules

Our bodies produce thousands of different molecular messages, and drug developers have taken advantage of this fact by synthesizing "impostor" messages that mimic the chemistry and shape of our natural molecules. In fact, many drugs used in medicine today achieve their results by preventing natural signals from engaging their receptors. For example, drugs known as beta blockers—frequently given to lower blood pressure—fit some of the adrenaline receptors and prevent our own adrenaline from sending its information. In so doing, they prevent or lessen the adrenaline's signal to "get ready for action" and keep the heart from racing.[3]

These imposter molecules can have wide-ranging effects. When I was getting ready to give my first talk to the American Heart Association, my boss, a cardiologist, asked whether I wanted to take propranolol, a beta blocker, to ease my nervous, racing heart; people who are terrified of speaking in public or taking exams sometimes resort to popping one of these before the stressful event. Though I chose to use natural methods—I meditated beforehand instead of medicating—it is fascinating that altering communication on a cellular level can both ease the physical condition of hypertension and powerfully impact our emotional experience.

Call for Rescue

The continual ebb and flow of molecular messages is essential to life and survival. When in danger, the cell calls for help, alerts its allies, and demands energy. Equipped with a chemical repertoire of molecular messages, it engages neighboring and distant cells in common action. When stress places demands on our energy, our cells can burn up their resources too quickly. When this happens, we can bring them replenishment and ease them back toward a state of peace through collaboration between mind, molecules, and cells.

They are incredible communicators, these molecules, speaking their myriad languages in simultaneous chorus within us. If you feel threatened by something in the present moment, or if a thought about the future worries you, your cells are called to immediate active duty to protect and defend your home turf, mobilizing resources to ensure survival. Moment to moment, we and our cells share the ability to move away from danger toward safe haven. And knowing that our cells listen to all messages—what we're thinking, imagining, or physically experiencing—gives us pause to remember how important it is to *be present in the now.*

Body Clues of Cellular Communication

Many years ago, Carl Jung, the Swiss psychiatrist who transformed the field of psychology, became aware that some emotional states were accompanied by corresponding physical reactions, such as a rapid heartbeat or sweaty palms.[4] Since then, many physiological signals have been harnessed into useful emotional measurements that can be detected by machines such as the polygraph (lie detector). Yet we've had machines for only a blip in human history. What about our own sensing abilities? Let's take some time now to become aware of our body's innate emotional signals.

The easiest molecular change to notice in our bodies comes from the cocktail of stress hormones. We fear a man walking down the street—he seems menacing. We narrowly avoid a traffic accident. We imagine awful things that will probably never come to pass: we worry. At the cellular level, adrenaline, the preeminent stress signal, prepares us for flight or fight.[5] Remember that it ensures that our blood sugar is elevated, which requires our heart to beat faster—a body clue that emotional and molecular change is taking place. Circulation shifts to fuel the big muscles in the legs, enabling us to run away if we have to. When blood moves to the legs, it moves away from the hands; this is why icy, clammy hands signal that we may be experiencing stress or fear. We breathe faster and shallower. The jaw clenches and muscles in the shoulder and neck tense. All these physical changes result from the communications between molecules and cells; in this case, molecules of adrenaline (along with other stress hormones) connect with receptors on heart, muscle, and lung cells—and in the case of long-term, sustained stress, immune cells.

When our cells broadcast a signal of danger, the whole body responds with detectable evidence. The same is true with the opposite signal: the all-clear that comes when we realize the "menacing" man is smiling hello as he passes by, or our near-accident has been avoided. We relax. Our breathing slows down; our clenched jaw and tight muscles release their tensions, and our hands warm up. Just as our cells listen to their surrounding environment, so can we listen to the echoes of their activity within us. And as our awareness of these responses increases, we can learn how to manage and, if need be, influence them intentionally. As we get better at reading our body's clues, we can learn to respond in healthier ways.

Set aside a few minutes now to tune in to how your body is feeling in this moment.

Exploration

Body Scan

Close your eyes. Notice whether there are any places in your body that feel tense or tight.

Notice the temperature of your hands.

Now put one hand on your chest and the other on your belly and become aware of the rhythm of your breathing. Which has more movement: your chest or belly?

Notice how you are holding your jaw and shoulders.

These observations give you a sensory picture of *now.*[6]

If you are inclined to carry this exploration further, remember a particularly stressful time or a frightening moment. Hold that memory in your mind and notice whether there are any physical changes from the earlier *now:* Are you breathing faster? Have your shoulders tensed? Let that memory go and imagine a restful scene or a peaceful moment to bring your cells back into balance.

◇◇

Your thoughts and your cells have just had a conversation about danger and safety. Appreciate the miracle this represents—the sacredness of the interplay of mind and molecules. Be aware that it is not simply physical events that trigger our "stress cocktails"; our minds play an important role in our chemistry.

Body clues can make you aware of your inner emotional state, and they are with you always—you can pay attention to them anywhere and at any time. For example, next time you're in a meeting and find that your hands are freezing, note that your cells may be saying, "Danger." Now that you have received their message, what action do you want to take?

Cells Speak Their Truth

In the late 1880s, a patient complained to her physician of electric tingling in her hands and feet whenever she caught a whiff of certain unpleasant odors. Her French physician was later to discover that electrical properties of the skin changed with fluctuations in emotion. Ultimately, from this discovery modern psychophysiology was born. Electrical activity of the skin became known as the galvanic skin response (GSR), and machines were developed that could measure it, such as the lie detector mentioned earlier. GSR, a measurement of sweat gland activity, is an index of events in the brain that are carried to the surface of the skin. Carl Jung, one of the first students of the skin's electrical response, viewed this as a physiologic window to the unconscious.

Imagine the excitement in those early days. You're sitting in a musty old lab, your fingers attached to an elaborate set of wires connected to a huge machine. Every time you imagine a friend's face, the needle on the machine moves. You realize that the friend doesn't have to be there for you to react emotionally and physically: she is present only in your imagination, yet she is changing you physically.

Fast-forward to the late 1980s when psychologist James Pennebaker was invited to teach the psychophysiology of stress to technicians who administered polygraph tests. They wanted to know what was occurring in the body and mind when a person was being questioned about a crime. Typically during the test, a person telling a lie reacts with a measurable stress response. In addition to the GSR and skin electrical conductance, modern polygraph measurements can include heart rate, muscle tension, voice changes, and other links to emotional discomfort. Pennebaker could explain the physiologic mechanism underlying each of these changes—yet he was to learn something that would transform his life's work and our knowledge about our cellular selves.

The revelation came when those experts administering the lie detector tests asked Pennebaker to explain this surprising observation: when a person actually confessed to a crime, they exhibited relaxation responses, not stress. As a result of their admission, they now faced a future of upheaval and turmoil, possibly even incarceration. How could they possibly respond by *relaxing?*

Pennebaker had no answer at the time, but he would later make a startling discovery that now informs what we know of personal well-being and telling the truth. Following up at the University of Texas with his psychology students, he began exploring confession itself. He asked his students to "confess," in writing, to a secret or a trauma they'd never told anyone. He discovered that following this disclosure, his students' immune health improved and their levels of stress hormones decreased.

He explained the phenomenon this way: When people hold back a painful or fearful story (a crime committed, an abuse suffered or inflicted on another, a secret fear), the very experience of holding back is stressful, and their cells respond accordingly with classic symptoms of discomfort and anxiety. When the story comes out, there is a wave of release and relief. In other words, revealing their well-guarded secrets enables them to let go of the associated upsetting thoughts and allows them to return to a state of well-being. Now their cells have the opportunity to initiate the chemistry of peace—and they do.[7]

For nearly twenty years now, Pennebaker has been giving people a simple writing assignment, which you will find in the next exploration exercise. Many who follow his suggestions find their immune systems strengthened. Students see their grades improve. Sometimes lives are completely changed. People of all kinds benefit: those grappling with everyday worries, those who have lost their jobs, people dealing with a terminal illness, victims of violent crime, even Holocaust survivors.

In his initial research, Pennebaker was most interested in health problems, so he turned to people with powerful secrets such as a history of abuse, who were more prone to illness. He wondered, if he could find a way for people to share those secrets, would their health improve? Yes, it turned out, it would, and it wasn't even necessary to reveal a word to anyone else. The simple act of writing about those secrets, even if the paper they were written on was immediately destroyed, had a positive effect on health. And it worked for people with an array of health challenges: people with asthma had better lung function; those with arthritis, less joint pain. A key finding of the study was that even though people wrote for only four days, fifteen to twenty minutes at a sitting, their symptoms continued to improve for the entire six months the study ran.[8] This is a potent tool in our medicine bag of cellular strategies, one that allows us to influence what we ask our cells to listen to.

Pennebaker's exciting research demonstrates that our cellular sanctuaries instinctively live by the New Testament declaration, "The truth will set you free."

◇◇

Exploration

You and Your Cells, Telling the Truth

You do not have to be suffering from a major trauma or illness to discover how this simple strategy might benefit you. Take fifteen to twenty minutes each day to write your thoughts and feelings about what is presently bothering you: your biggest stressor or hidden shame. Commit to writing for four consecutive days—no more or less. Don't write the facts about the story; rather, express your emotions about them. It's not necessary to reread what you have written or show it to anyone. You can tear it up or burn it if you like. This is an exercise for you and your cells, an invitation to release and let go.

◇◇

There is a profound lesson within the dynamics at work in our inner sanctuary: our cells know the truth. Our physiology responds to what we're thinking, including what we don't want people to know. We can hide something from friends and family, our neighbors, and coworkers. We cannot hide from our cells. They are listening to all the conversations in our head, every word we whisper to ourselves. And because of this, when we release the stories and feelings that torment us, our cells respond with great relief. They become havens of safety once again.

Our Cells Listen—Do We?

Because of what I have learned about our cells, it is plain to me that they listen to what we are thinking and respond accordingly. This has prompted me to watch my own listening behavior, which recently led to a fresh insight—see if you recognize this in yourself. I'm walking with a friend who's telling me a story I've heard before, more than once. Internally, I react to her words and start judging what she's telling me. Then—an "aha!" moment: a little inner voice asks, *How about just listening to her now instead of reacting? What would happen then?* I am prone to reacting, at least internally, to what people say to me. In fact, I can send my entire physiology into doldrums or wars depending on what I hear. What about you? Just as when we hold in secrets, our cells respond when we have an internal conversation judging or criticizing someone; they hear that too and react with stress or anger. So there is no need for our inner critic to do anything other than disappear. Being reminded of *now* by my inner voice, I simply listened to my friend; my body relaxed, and I felt more deeply connected to just being with her. What a difference it can be to listen without the internal judge being turned on.

Since that epiphany, whenever I find myself beginning to react while listening, I remember my cells and recall that I have a choice to simply listen. I feel better as a result, and so do my relationships. This is not to say that I don't interact in a conversation; I am simply alert for a shift into reactive mode, and I consciously bring my awareness back to hearing what is being said. I am learning to consider my cells: they are listening to my inner conversations as well as to the external world, and I can choose not to feed them unnecessary information, such as negatively judging a friend internally or even criticizing myself.

This lesson from our cells holds wisdom for us. We all need people to listen without judgment as cells do: our loved ones, our coworkers, even our governments. It is important for our survival and our well-being.

The Company We Keep

So now we have seen that our cells are in relationship with our thoughts, feelings, and each other. How do they factor into our relationships with others? Listening and communicating clearly play an important part in healthy relationships. Can relationships play an essential role in our own health? More than fifty years ago there was a seminal finding when the social and health habits of more than 4,500 men and women were followed for a period of ten years. This epidemiological study led researchers to a groundbreaking discovery: people who had few or no social contacts died earlier than those who lived richer social lives. Social connections, we learned, had a profound influence on physical health.[9]

Further evidence for this fascinating finding came from the town of Roseto, Pennsylvania. Epidemiologists were interested in Roseto because of its extremely low rate of coronary artery disease and death caused by heart disease compared to the rest of the United States. What were the town's residents doing differently that protected them from the number one killer in the United States?

On close examination, it seemed to defy common sense: health nuts, these townspeople were not. They didn't get much exercise, many were overweight, they smoked, and they relished high-fat diets. They had all the risk factors for heart disease. Their health secret, effective despite questionable lifestyle choices, turned out to be strong communal, cultural, and familial ties.

A few years later, as the younger generation started leaving town, they faced a rude awakening. Even when they had improved their health behaviors—stopped smoking, started exercising, changed their diets—their rate of heart disease rose dramatically. Why? Because they had lost the extraordinarily close connection they enjoyed with neighbors and family.[10] From studies such as these, we learn that social isolation is almost as great a precursor of heart disease as elevated cholesterol or smoking. People connection is as important as cellular connections.

Since the initial large population studies, scientists in the field of psychoneuroimmunology have demonstrated that having a support system helps in recovery from illness, prevention of viral infections, and maintaining healthier hearts.[11] For example, in the 1990s researchers began laboratory studies with healthy volunteers to uncover biological links to social and psychological behavior. Infected experimentally with cold viruses, volunteers were kept in isolation and monitored for symptoms and evidence of infection. All showed immunological evidence of a viral infection, yet only some developed symptoms of a cold. Guess which ones got sick: those who reported the most stress and the fewest social interactions in their "real life" outside the lab setting.[12]

We Share the Single Cell's Fate

Community is part of our healing network, all the way down to the level of our cells. A single cell left alone in a petri dish will not survive. In fact, cells actually program themselves to die if they are isolated! Neurons in the developing brain that fail to connect to other cells also program themselves to die—more evidence of the life-saving need for connection; no cell thrives alone.

What we see in the microcosm is reflected in the larger organism: just as our cells need to stay connected to stay alive, we, too, need regular contact with family, friends, and community. Personal relationships nourish our cells, ourselves, and our souls. I can well imagine that a physician of the future might write this prescription on her pad: "Take two friends out for dinner or a walk and call me in the morning." But we don't need to wait for that enlightened doctor; we can prescribe this for ourselves. Let's get past vague promises and best intentions: let's *really* do lunch.

> *The great sage Hillel taught: "If not now, when?" There is only one now. The past is a memory and the future is a dream. . . . Only the now is real—make each second of your life count.*
>
> —RABBI DAVID AARON *The God-Powered Life*

The Cell and Now

Of course, the *quality* of our communication and relationships also influences our cells. Fascinating studies of newlyweds by Janice Kiecolt-Glaser and her husband Ronald Glaser at Ohio State University give us a look at how our cells are influenced by the nature of how we deal with conflict. Newlywed couples were monitored over a twenty-four-hour period during and after they had a heated disagreement. Throughout this time, their blood was evaluated for changes in stress hormones.

It was not surprising that researchers found elevated stress hormones in both men and women during and right after the argument. What was more remarkable was that the disagreement caused different physiological and psychological responses depending on gender. The women's stress hormones remained elevated for some time afterward, while in general, after the argument was over, the men tuned it down—and tuned out their wives. The women continued to relive the argument, and to make matters worse, now they were angry at being tuned out too. This study is a great reminder to let go of a stressful situation once we've experienced it in "real time." If we relive it over and over, we put our bodies, and their trillions of cell sanctuaries, in great distress. We risk illness and fatigue when our cells are thrown out of balance in this way. And we have a choice.

While our cells are always in the now, our thoughts can sentence them to reliving *yesterday's* now. As the mind reacts, our vigilant, ever-listening cells respond to its call. That is why it is important to nudge the inner kvetch into a more peaceful state.

The next time your mind is sweeping you off to stressland, consider your cells as the sacred vessels of your life, and give them the present they deserve. Shift your attention from past or future, leave the realm of "Why did . . . ?" and "What if . . . ?" and let your mind rest in what is right in front of you, around you, within you. Be a friend to your cells. Bathe them in the molecules of peacefulness and contentment, in the chemistry of love. Now.

A few years ago I was driving to the California town of Sonoma to give one of my Cells and the Sacred workshops at the Science Buzz Cafe. Along the way, my mind was jumping back and forth between money worries and boyfriend angst. Instead of enjoying the blissful vineyard scenery, my awareness was elsewhere. And then, "aha!" A thought leapt into my consciousness: *Your cells are* always *in the now.*

Message received, loud and clear. I was creating my own, unnecessary discomfort and tension, mentally barreling down a road to nowhere I wanted to be.

Wow, thank you, cells! *Now.*

Love Messages—Molecules That Bind Us

We have looked at the effects in our cells and bodies of danger, threat, and stress. Yet we have an internal pharmacy that can also deliver molecules of love and connection, empathy and relaxation. In fact, love and affection form the perfect bridge from science to the sacred.

We have investigated a major stress hormone, adrenaline, to illustrate how our cell receptors work. Yet during a stressful event, a slew of other chemical potions are released to ensure our survival. And this points to an interesting molecular difference between male and female responses to stress.

In addition to our cells releasing stimulating molecules such as adrenaline and cortisol, during stress both men and women release another molecular medicine: oxytocin, a bonding and calming molecule.[13] A hormone and neuropeptide produced in the brain and pituitary gland, oxytocin is typically associated with pregnancy and childbirth; it increases at the end of a woman's pregnancy to promote labor, uterine contractions, and lactation. A synthetic form of the molecule, Pitocin, is often used to induce labor, and in animals it initiates maternal behaviors. Distressing as it is to contemplate, a virgin female rat will actually cannibalize newborns of her species unless given a dose of oxytocin. Receiving oxytocin brings about an abrupt change of heart—she begins to behave like a kind mother rat and will now care for the babies. Oxytocin is the molecule that initiates bonding between mom and baby; in fact, it is the molecular message that initiates bonding behaviors in all warm-blooded animals.

Studies at UCLA showed that in response to stress, women typically "tend and befriend," and researchers provided a molecular explanation for this bonding behavior: oxytocin.[14] It was once thought to be only a female hormone, but we now know that both men and women make it—though its calming influence is amplified by estrogen and hampered by androgens, the male hormones.

Consider that in our earlier tribal days, men went off to hunt while women guarded the homestead, cared for the children, and attended to the creature comforts of the community: preparing food, maintaining structures used for shelter, and fashioning clothing and other useful items. Sharing these considerable responsibilities with other women ensured the survival of the entire group.

The cellular impetus for women to share remains in the present day. When women are going through a stressful time, their urge is to protect and nurture, tend and befriend: bake cookies, feed themselves, attend to their children's needs, call someone they trust. The female seeks to bond, wants to be heard and held by the people in her life. For example, during the newlywed fights researchers studied, if the husband had tuned back in to his wife, her stress response might have quickly tapered off. The same might have happened if she had the opportunity to call a friend. In contrast, though men also turn to buddy groups, in times of stress they tend not to be as upfront about their need to connect as women are.

This tend-and-befriend response goes back even further than human hunter-gatherers; it evolved from female primates' genetic programming. Male primates, with their different molecular profile, are often competitive and combative, withdrawing to take care of their wounds—physical or emotional. Psychologists think this basic biochemistry explains why men withdraw into solitary silence in response to stress or anger. They go to their man cave; women go to their kitchens or cell phones.

In oxytocin we have a molecular messenger, a mood changer, that can calm us down and promote closeness with other people. And it is a simple fact of gender that these molecules are more easily "managed" by the female of the species because of her abundance of estrogen.

Exploration

Grab a Dose of Oxytocin

Explore the quickest, easiest way to receive an immediate, welcome dose of oxytocin: hug someone. Give your sacred cells the gift of connection.

Enhanced Bonding and the Oxytocin Effect

The release of oxytocin into the blood can be initiated by kinship, connection, and physical touch. A newborn nuzzles his mama, recognizing her smell; she in turn nourishes him, and her touching and stroking encourage him to bond with her. And so it is with adults. Touch your friend's hand, massage your lover's shoulders; you are stimulating molecular messages that enhance emotional bonding.

A lover's sensual touch heightens oxytocin production, and this explains why foreplay enhances sexual arousal. Levels of oxytocin increase during sexual stimulation and peak during orgasm. In men, after ejaculation, high levels of oxytocin are maintained for about a half hour; in women, the levels drop sharply shortly after orgasm.[15] Maybe longer-lasting post-orgasm oxytocin explains why men often ease right into sleep after sex, as one of the effects of oxytocin is deep relaxation. Another effect is fuzzy thinking; if you feel "spaced out" after lovemaking, this may be what's going on—oxytocin temporarily reduces the capacity to think and reason.

Production of oxytocin increases with age; though women may lose childbearing abilities, as time passes the capacity for friending and bonding expands. Men involved in caring for their children also have elevated oxytocin levels and bond better with their babies.[16] When Dad holds his child, his testosterone level drops, in part because his "love chemicals" increase. Fortunately for us, there are many other "love chemicals": we have a pleasure pharmacy at our fingertips—literally.

"Better Bonding" through Chemistry?

Oxytocin may be the primary biochemical basis for human attachment and for achieving close emotional relationships. Warmth, imagery, massage, and hypnosis are some ways to spur connection. Drugs that encourage closeness include ecstasy and marijuana. Oxytocin may be present in a number of plants, including *Cannabis sativa* and blue cohosh root, or these plants may stimulate oxytocin release from our cells.

Bonding—like the perfect fit of a cellular hug between matched molecules—promotes health in many ways. Both blood pressure and stress hormones decrease in animals that receive injections of oxytocin. Their wounds heal better and faster. Best of all, these benefits last for several weeks after the final injection. Rabbits that were petted did not succumb to experimental bacterial infections that sickened their unstroked bunny relatives. This helps us understand why love, lovemaking, pets, and petting can promote health and relieve stress—and why close relationships can protect us from the harmful effects of stress. At present, the female of our species lives longer than the male—but might it be healthier bonding that prolongs our lives?

Like our cells, we bond with those we trust, know, and respect. Our cells show us the way: if they don't trust the information they receive from an invader or stress molecule, they marshal their forces to eliminate it. When, where, and from whom do you receive love and connection? Each time you make contact, you are sending molecules of love to your cells. And isn't our sacred nature imbued with love?

The Bungee Effect in Relationships

I have recently begun to understand the "bungee effect," and I will share a personal story to illustrate it. For decades I had been in a long-distance love relationship. His scent intoxicated me, his touch enthralled me, and for twenty-four hours each time we met, we couldn't get enough of each other—body and soul. I'd gotten used to this bizarre arrangement, and though I longed for something more, it had me hooked. Each time we separated and he disappeared "into the sunset" of a thousand miles away, I got on with my life—the bungee stretched to its full extension. Then, after a while—usually around three months—I felt a tug on the cord. I began to think about him, wonder if he was OK, and then I needed to know. I needed another "hit": a voice, a touch, a connection. He had similar experiences while living out there at the other end of our tether, content with no contact—until . . .

I was perplexed about this pattern for many years before I made the mental connection: oxytocin. When we came together we gave each other massive doses of cellular "medicine"—love molecules—enough to get us through the next few months content and at peace. Then the dosage wore off and *snap* went the bungee cord. We needed replenishment, and we had "the urge to merge" and fuel one another's bonding molecules once more. Intimacy connects people at a cellular level, a bond that is not easily broken for good.

Prayer Receptors?

We have learned a great deal about how our cells listen for molecular messages, but what about other, less measurable communications such as intention, spiritual guidance, and prayer?

While a resident at San Francisco General Hospital, cardiologist Randy Byrd laid the framework for groundbreaking studies on the power of prayer by exploring whether prayer would help people in the ICU who had suffered heart attacks. Though the patients in the study did not know they were being prayed for, they had fewer second cardiac episodes and complications than those who weren't prayed for.[17] Since then, numerous studies have arrived at similar as well as contradictory conclusions: being on the receiving end of prayers may improve physical health.[18] Our cells must be able to get the message somehow. Where are our cells' receptors for prayer? So far, none have been revealed. Scientists such as Larry Dossey, another pioneer in the power of nonlocal healing, have speculated that prayer energy travels as extremely low-frequency (elf) electromagnetic waves that somehow touch and influence our cells.[19] Though it is a mystery at present, here we find another form of cellular call and response.

Loving Our Cells, Our Selves

The sacred messages from this chapter are to regard our cells' ability to listen and receive with gratitude and reverence and to consider the ways in which we do this in the fullness of our lives. Let's pay attention to the words we say to ourselves and others. Let's watch the ways we reach out to connect. And remember: our cells are listening.

Our cells are formed in the intimate embrace of molecules receiving from God the holy spark of life. They are sacred vessels for divine love seeking connection in the present moment. While cells are real and solid—observable entities that make us who we are physically—they are also reflections of mystical teachings through the ages. Contained within them are lessons about giving and receiving, opening and softening, nurturing and protecting, community and truth.

Our cells can only receive when they are open to recognizing the information offered to them. Let's ask ourselves: How like our cells can we be? Are we open to receiving what fits best? Do we give to others, as our cells give their information freely? Do we offer messages that masquerade as truth, or do we share ourselves with integrity? Are we living in the *now?*

Our cells hold the answers to all these questions. Let's remember to listen.

Reflection

How receptive am I?

How deeply do I listen?

How well do I communicate?

What and how do I communicate best—words, emotions, prayer, energy fields?

What do I need to communicate?

Do I speak the truth, live the truth?

Does my usual communication foster cooperation?

How do I hold my relationships as sacred?

What are my most sacred relationships?

◇◇◇

We all have much more listening to do.

—MARY OLIVER

Chapter 4

The Fabric of Life–Choose

Each cell can take in information about its circumstances and respond to it purposefully.

—BOYCE RENSBERGER *Life Itself*

What is it that tugs on the edges of our cells and consciousness and urges change? What convinces the cell to choose one focus of attention over another?

Cell biologists have long believed that a cell behaves the way it does because of genes, proteins, and signaling molecules. Yet pioneering scientists now show that by physically twisting, bending, and pushing the cells, *mechanical forces* help control which action a cell performs.[1]

Embedded in the design of our cells is a translucent, dynamic webbing that decides the cell's direction. While the external receptors we have learned about in preceding chapters *listen* to our molecules, the fabric or "strings" of our cells *manifest action.*

Figure 4.1 Spiderweb illustrating a similar design as the cellular cytoskeleton

Connecting inside with outside, the strings vibrate, push, and pull, guiding the cell into delivering what it's supposed to. A new fluttering on its strings plays a new tune of activities. This is the way into the secrets of our cells.

The degree of tension on the matrix of the cell regulates the cell's expression and destiny.[2] Stretching taut triggers one genetic message and outcome; letting go of some tension initiates another message and outcome. Same genes, same internal intelligence—different future. This process of balancing forces and tension is a universal law of design called *tensegrity.*[3] Tensegrity guides the pattern of human-made structures, cells, and even complex tissues. We find it in buildings and atoms, spiderwebs, stars, and molecules (see figures 4.1–4.3).

Definition

Tensegrity: Refers to any physical structure that stabilizes and supports itself by balancing opposing forces of tension and contraction. Structures are stabilized mechanically by balancing internal and external forces.

Figure 4.2 Buckminster Fuller's geodesic dome, Toronto, Canada

The term was constructed from "tensional integrity" by architect-futurist Buckminster Fuller to describe situations in which push and pull have a win-win relationship.[4] Bucky used it to build his famous geodesic domes, the most stable of human-built structures (see figure 4.2).

◇◇

Donald Ingber, when taking a design class as an undergraduate biology student in the 1970s, learned about sculptures that relied on tension to hold long tubes together and create stable forms. As he contemplated this, he had an intuition that cells, too, must be tensegrity structures. Now a Harvard professor, Dr. Ingber has put tensegrity on the map of cellular design, regulation, and intelligence. At a biological level, tensegrity allows us to comprehend how changes in shape and mechanical strain influence cellular choices and actions.

Had I not experienced a moment of synchronicity, I might have overlooked this important aspect of the cell altogether. It was 1998, decades after I had studied biology, when I was in a bookstore perusing popular magazines. Two articles at opposite ends of the shelf attracted

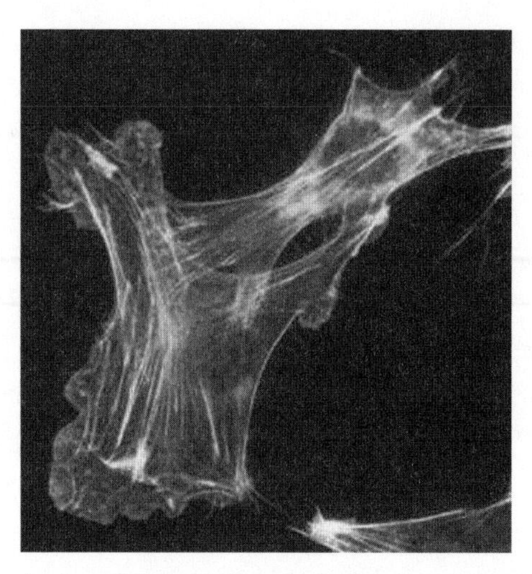

Figure 4.3 Tensegrity (cytoskeleton) in actual cell – (mouse embryonic fibroblast line) represented by the long thread-like structures; the dark round in the upper right is the nucleus; image by Feldman, M. E., et al.

my attention: one in the *Yoga Journal* and the other in *Scientific American.* Both used the Fuller-coined term *tensegrity,* which I had never heard before. One of the articles, written by Carlos Castaneda, concerned ancient practices he referred to as *tensegrity movements,* which were said to alter human consciousness.[5] The other, by Dr. Ingber, delved into the very architecture of life. He described the cell as having a tensegrity structure that guides its decision-making abilities. The notion that this architectural principle could be at work in both the microscopic stuff of which our bodies are made and in our consciousness came as a revelation.

The Architecture of Life—Cellular Mastermind

This remarkable architectural design as it manifests in living cells is the *cytoskeleton.* Likened to the cell's muscle and bones, the cytoskeleton is the scaffolding that connects all parts of the cell. It also prevents the cell from collapsing on itself. This cytoskelton matrix transports

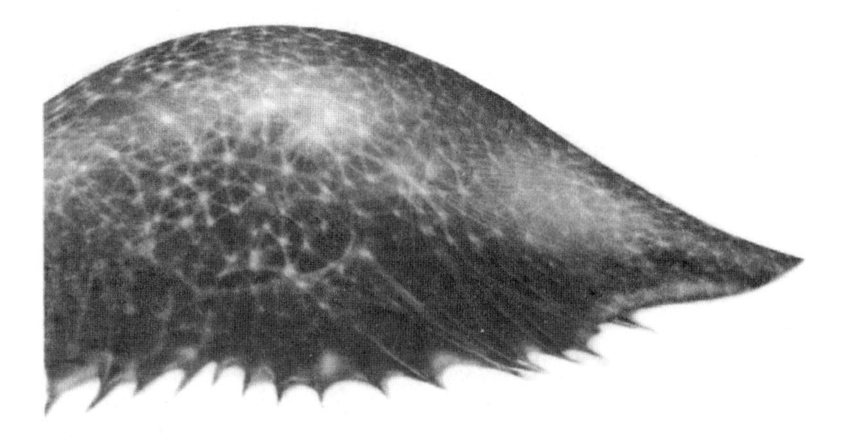

Figure 4.4 A drawing of the cytoskeleton fabric; image by Slim Films

molecules, coordinates information, and regulates genetic expression. With the ability to balance the push-pull of the cell, it is the newest biological candidate for the seat of cellular intelligence as well as the seat of consciousness.[6]

Many scientists still contend that the cell's intelligence is housed in its genes. Yet genetic intelligence is simply a vast text of chemical codes constructed from long, spiraling molecules of DNA. The text provides recipes for making the necessary protein ingredients for life—yet who, and where, is the "cook"? Some critical thinkers maintain that we could get closer to the cook—and our dynamic cellular intelligence—if we investigated how our cells are built instead of deciphering their genes. Put another way, in cellular communities, our genes are the plans; the cytoskeleton is the mastermind.

Noted scientist Dr. Bruce Lipton took cellular intelligence to the next level—beyond our genes to the receptors on the cell "mem-brain." Here we discover that cellular intelligence is carried in the interplay between receptors and the cytoskeleton.

Let's imagine shrinking ourselves until we are tinier than the cell itself so we can examine this fabric and inner scaffolding; Nobel Prize winner Christian de Duve would call us cytonauts—sailors inside the cell. To enter the sanctuary of the cell, we must first sail past the quivering receptors on the outside surface. Once inside, the sound of thousands of miniscule maneuverings attracts our attention. As we pause to listen, we observe the large "heart," or nucleus, at the cell's center. We may even hear the humming of hardworking energy generators, the mitochondria. Experimenting with the surface beneath our feet, we begin to gently bounce as if on a trampoline. Below us we see nothing but the translucent, gelatin-like cytoplasm. Looking closer, we notice tiny shimmering strings and tubes throughout this "Jell-O" holding us up and reaching throughout the cell. When we move, the strings respond. If we bounce or step lightly on one part of the cell, the rest of the cell adjusts to the change in tension. This dynamic, vibrating matrix is the true mastermind of cell intelligence.

The cytoskeletal fabric is composed of three different kinds of organized proteins: fat tubes (microtubules); skinny microfilaments; and long, willowy intermediate filaments. Acting as struts and pulleys, these vibrating filaments, tubes, and strings permeate the cell like a web. Each exerts the power to direct, manage, and coordinate cellular behavior.

Scientists have known for decades that microtubules help cells move, change shape, and divide, but only recently have we learned that they are also partners in managing cellular tension. And *a change in tension affects genetic expression* and, hence, cell abilities. Simply said, *alterations in the cell's physical state can alter its genes.* A cell that is stretched out, for example, has a different fate from one that is balled up—even though they contain the same genetic information. When pulled, pushed, plucked, or released, cellular scaffolding manifests different abilities and genetic programs. This dynamic interplay of forces keeps the cell "listening" to "choose" what to do next.

Signals felt like a pebble dropping on the surface of a pond, waves send responses inside the cell, so that the message can be heard.

Then silence, waiting for action or further listening.

—CHRISTOPHER VAUGHAN *How Life Begins*

Shifting Attention

Let's consider this further. Cells change shape and tension. They may stiffen or relax, and each physical state affects what the cell can do. For example, when an immune scavenger cell receives a tug—let's say, a message of bacterial invasion—it responds instantly. Elongating its usual spherical shape, it moves deliberately toward its prey. Upon meeting the invader, the cell attaches to it with sticky proteins, changing shape again to wrap around the intruder to eliminate it. This response requires the membrane receptors to recognize danger (that is, "not self") and attach, while the fabric inside the cell responds and coordinates the cell's activities.

Another shape changer, the microtubules, continually dismantle and rebuild about every ten minutes so that our cells are in a constant state of change and readiness, rebuilding to respond.[7] This also shows how flexible we and our cells are in allowing change. (See plate 1 in the color insert for a cell photograph of a human white blood cell (phagocytic neutrophil) recognizing and going after red blood cell from another species.)

Cellular Decision Making: Life and Death

Cell tension and shape orchestrate life and death.

Living cells do "either/or": they *either* reproduce *or* they mature; they make copies of themselves or "grow up." A reproducing cell exercises genetic intelligence only to make another cell—it doesn't produce the resources a mature cell needs. The mature cell engages a different set of genes to provide for its livelihood, and it does not reproduce (see table 4.1). The hidden strings and fabric of the cytoskeleton regulate all of this.

When we look at how cells establish growth in petri dishes, we see them anchor firmly onto the surface and then spread out on their new plastic home. Cells attach, stretch, and spread out, signaling the genes to begin dividing. New cells are made until the entire surface is covered: just as skin cells cover a gash to heal the wound, cells stretch to make more cells.

If the plastic home gets too crowded, some of the stretched cells let go of the dish and begin to assume more rounded forms. If too many cells are competing for the same space and resources, genes that trigger duplication are turned off. Cells may even turn on self-destruct genes. Consider that cells "choose" to sacrifice themselves for the benefit of the community, to make more room and food available for the rest. Is this a sign of cellular altruism?

Between the extremes of being stretched out with maximum tension (signaling growth genes) or balled up with little tension (signaling death genes), there are cells whose tension is "just right." In this in-between physical state, genetic programs instruct cells to manufacture the special "goodies" of fully developed, mature cells. Mature genetic responses maintain what's needed for their—and our—lives. Cells reproduce, mature, or die, one state at a time.

Table 4.1 Examples of Reproducing and Maturing Cells

Reproductive Phase: Make More Cells	Maturity: Functions and Products
A liver cell divides, making more liver cells	Detoxifies drugs and poisons with newly made proteins
An adrenal cell makes copies of itself	Manufactures adrenaline and releases it into the blood
An immune cell produces more lymphocytes	Produces antibodies, interferon, and protective potions

All cells in the body carry the same genes. (The exception is mature human red blood cells, which have no genes.) One gene program is launched by a tug or pull on the cell, while another is launched when tension is released. The genetic instructions carried out are influenced by cellular tension, location, and the body's chemical cocktails. The cytoskeleton is behind all of this cellular decision making.

Cells attached and stretched to their limits make copies of themselves. Over and over again, they repeat the same pattern. Cells that let go of their sticky attachments yet still retain flexibility and strength exert their maturation powers. Cells that do neither, completely let go—release—round in on themselves, and die a gentle death.

Cellular Buddhism?

Learn to let go and allow the changing mystery of life to move through us without fearing it, without holding or grasping. . . . Letting go and moving through life from one change to another brings the maturing of our spiritual being. In the end we discover that to love and to let go can be the same thing. . . . Both allow us to touch each moment of this changing life and allow us to be there fully for what arises next.

—JACK KORNFIELD *A Path with Heart*

According to a Buddhist concept, when we learn to let go of attachments, we mature on our spiritual journey. So it is with our cells. When they let go some—but not too much—they evolve into more mature citizens. They are farther along in their life's journey.

Could the idea of letting go of attachments to achieve spiritual maturity and enlightenment have arisen only from philosophical and psychological tenets? Or did clues for this idea originate with an observation or vision of the internal microscopic universe? What influenced the spiritual concept of maturity following a letting-go process?

Reflection

Attachments and Letting Go

A seed of the mighty oak tree nestles in the womb of Mother Earth.

Sending down shoots and roots, it attaches to the Earth Mother.

Only when it's connected and attached can it grow upward toward the light.

A single, tiny fertilized egg, anchoring into the mother's womb, attaches and proliferates into a trillion-celled baby. Letting go of the womb into the bright light of the world, the baby matures.

Throughout life we go through patterns of repeating ourselves, growing and maturing until our very last breath, when we fully let go.

Exploration

The Three States

In teaching, I have discovered that when we embody cell behavior, we learn concepts physically through our bodies that may be difficult to grasp with our minds. The following exploration of cellular states helps engage your cellular wisdom. Give yourself at least fifteen minutes for this exercise. Have ready a pad of paper and a special pen or a set of colored pencils, or anything else you might want to use to write about your experience afterward. If you are in a group, one person can be a timekeeper, calling out when to change states at about five-minute intervals or longer. If you are doing this by yourself, you can sense when it's time for you to change or set a timer.

Sit comfortably on the floor and let your mind and space be free of any distractions. Imagine yourself to be a cell that has choices to make. Do you want to be one that is attached, stretched out as far as you can reach and clinging to the floor, making more cells, repeating yourself? Next you might choose to experience being all balled up, surrendering to a gentle dying process. Or you could become a maturing cell, no longer tightly attached to the floor and now gaining in maturity; you can do this acting as any kind of fully developed cell, such as one of the immune cells, a dancing neuron, or a beating heart cell. Use your imagination and pay attention to which state you want to enact first. Be with it for at least five minutes. Then shift to another state, feeling and sensing what that's like. Finally, choose the third state. You can do this in any order. When you feel the exercise is complete, write down what you experienced. What did you become aware of and what did you learn?

This is always a fascinating exploration during which people learn a great deal about themselves—it may be even more important than a better understanding of biology. Consider these states as metaphors for your own behavior. What were you drawn to first? Do you need to be making more of yourself, repeating some behaviors? Are you stuck in the same repetitive patterns, holding on too tight and not able to let go? Is there something you must fully let go of? Are you in a maturing phase of your life?

◇◇

One of my students said that before doing this exercise she did not want to experience the death phase. So she put it off until almost the last minute. When she finally allowed herself to experience fully letting go, she felt great relief. It wasn't scary at all.

Letting Go—What If . . .

Until the summer of 2010, I saw the letting go process mainly in cellular, psychological, and metaphorical terms. Then I met someone who led support groups for people with cancer. In our discussions about cells and healing, a new possibility in letting go arose.[8]

He told me about people he's worked with whose cancer healed or went into remission when it "shouldn't have." The common factor he observed was that they'd let go of something *big.* Maybe it was their fear of dying, perhaps a detrimental relationship: a *big* letting go. He included himself in that group, now six years in remission after a deadly diagnosis of advanced stage IV lymphoma.

I reflected with him on the cell's letting-go process, and we both made huge discoveries and insights. Let's revisit that conversation now.

Cells attached with extreme tension keep making the same cells. The more attached and tense we are about something, the more we make the same thing: the same mistakes, the same program, body, mind, and cell. Unable to get free, we are forced to repeat the same old stuff.

My extreme mental tension about *not*—not having enough, not being good enough, the whole knot of "nots"—may hold me attached to that position. What if I finally let go of "not enough"? Will my body and mind allow me to receive a different message? Would moving my body through the different cellular states help me release this old pattern?

When we fully let go of attachment, of our "stuck-ness," perhaps we enable our cells to let go of a program they no longer need to carry out. If we fully let go, is it just possible that the cells we no longer need, like cancer cells, can now program themselves for death? Is it possible that people with cancer who let go *big* give their cancer cells permission to let go, effectively initiating the cells' own death?

What if a more intimate understanding of our cells gives us a deeper understanding of ourselves? What if letting go of a fear or a relationship or a destructive situation is not simply about emotional changes but exists concurrently on a cellular level? Just what if . . .

The questions we pursued in this conversation, which I have reflected on ever since, do not lead to the conclusion that a person's cancer remains because he or she doesn't/won't/can't let go: cancer is far more complex than that. These questions simply—or not so simply—offer a compelling idea about healing.

Doorways to the Unknown: About Cancer

Is it possible to change cancer cells by changing their tension, or their environment? Scientists in the lab are beginning to discover that some cancer cells are more rigid than normal, healthy cells, and that rigidity or stiffening triggers disorganization away from normal cell growth.[9] When I first read that mammary tumor cells were stiffer than normal breast cells, I was not surprised, since many cancer cells are less mature and thus less flexible than normal cells. In general, normal immature cells are more rigid than mature cells. University of Pennsylvania scientist Dr. Valerie Weaver, examining cells grown in tissue culture, discovered rigidity in cancerous mammary tissue relative to healthy tissue. Not only that, she demonstrated that healthy cells that were grown on stiff materials showed abnormal tissue organization. Why? Because their cell "strings" were being pulled, increasing the mechanical tension.[10]

By contrast, chemicals that prevented pulling, attaching, and tensing permitted cells to grow into more normal-looking mammary tissue. Weaver hypothesized that mutant genes could activate biochemical pathways that increased cell tension, an early event leading to cancer. Furthermore, she suggested that by interfering with the mechanics of a cancer cell, we may be able to override aberrations in its genes. Researcher Donald Ingber is also studying whether changing the physical environment of a tumor can reverse the cancer process; he's attempting to manipulate a tumor's microenvironment by implanting artificial materials that mimic a healthy matrix. Will giving cells a softer "mattress" upon which to grow provide an environment in which cancer cells can change their genes?

We are always trying to solve the riddle of miraculous healings. Could one explanation be that when people soften their attitudes and tissues in some way, this triggers the cells to move toward normalcy? If I hadn't spent years in the lab showing that leukemic white blood cells could be pushed toward becoming more normal, it might not occur to me to ask this question: *can* cancer cells become normal? We accomplished that change in the lab using benign chemicals, though we didn't test whether those molecules worked by changing the tension of the cells or the genes. However, we did learn that their membranes became more fluid as they picked up more mature abilities in the process. But outside the lab, the question remains.

And it is followed by other questions that widen the scope of our discussion: If rigid environments can contribute to the disorganization of normal cells, what about rigidity when it comes to individual people or whole societies? How does inflexibility influence our development? What might we each be able to do to create a softer, gentler environment?

Ever wonder why someone doesn't try softer?

—LILY TOMLIN

Looking back into the microscope, we find scientists discovering that if embryonic stem cells are grown *in vitro* on a stiff structure, they will be more likely to become muscle cells; when grown on a soft, rubbery structure, they will become neurons.[11] In other words, the mechanics of the environment influences the tension in the cell and its genetic expression, telling the stem cell what it is to become and which genes to use.

Changing Cancer

My passion about cancer prevention and treatment came from spending a great deal of time with children threatened by the disease. I saw the hopeful yet devastating effects of chemotherapy and radiation and questioned whether there could be something else, some other less-toxic strategy, to reverse the disease. My explorations ultimately brought me to learn more about our cells and the role of tensegrity in physically altering a cell's outcome. It also brought me to ever-deepening shamanic and inner practices for healing.

Crossing Over

In the preface I related my experience as the "balloon lady" on the pediatric floor of the hospital and of my great friendship with little Alvaro. When his leukemia returned after a year of remission, I was overwhelmed—his death seemed inevitable, and I didn't know what to do. I called on the support of Dr. Tomas Pinkson, a psychologist who was clinical director for the Center for Attitudinal Healing at the time. Tom had also started one of the first hospices in the United States, so I knew he dealt with the end of life every day. When I called to ask him how to deal with Alvaro's impending death, he told me, "You don't deal, you feel." Of course, I didn't *want* to feel—that's why I was calling him. I was sure I'd be overcome with grief.

I met Tom in his office. Expecting a typical pristine professional space, I was surprised when he led me into what seemed to be an indigenous healer's space: it was entirely another world. On the floor were a Navajo rug and drums, on the walls were drawings and Huichol art. Tom invited me to sit on the floor across from him. Once I was settled, he lit what looked like a small bundle of twigs, blowing on them to raise smoke. Then he waved the smoke all around me. I would later learn that burning sage ("smudging," as it is called) is a Native American practice for cleansing and focusing energy.

From that moment, Tom's nonintellectual, nonlinear "shamanic" approach to inner wisdom intrigued me, and he became my lifelong teacher of healing—heart, mind, and soul. From him I learned a bit about letting go of my questioning mind and reaching into my knowing heart.

Changing Consciousness

My first experience of shamanic consciousness takes place at a drumming circle on the "left coast" of California, a place geographically, psychologically, and spiritually on the leading edge of human development. The year is 1985, and the place is a funky community center. I have traveled narrow, winding country roads to arrive here at Tomas's invitation.

Flickering candlelight brightens the darkened room. People are sitting on the floor in a circle. Some are drumming, shaking rattles, and chanting, and some are still: it is a strange scene to my scientist's eyes. I wonder why in the world I agreed to come here. I sit down in the open space nearest the door—in case a quick getaway is called for. My mind chatters endlessly: *Get me out of here* now. *You're crazy to be doing this, and so are all the people sitting in this circle!*

I close my eyes, believing this will help me endure what I think will be a huge and awkward waste of time. But then the sound and vibration of rhythmic drumming and rattling begin to nudge me to a quiet place inside. Despite my mental struggle and the physical squirming that accompanies it, something else takes me over. The chatter stops. Images begin to fill my mindscape. And then the reason I am here becomes clear: I am to connect with a deeper kind of wisdom than I have experienced before, to seek a knowing beyond my intellect. Soon I am being held in this sacred space, and a sense of calm envelops me. I feel connected to myself, to the other people in the room, and to the divine holiness of the moment. Moments of inner peace, tranquility, and—who knows?—maybe even enlightenment penetrate my being.

Afterward, I couldn't deny my surprising and almost reluctant experience of the sacred. But my chattering mind resumed its monologue—how to make sense of this? I needed a physical explanation of how changes in consciousness and emotions occur before I could truly trust what had happened as genuine. I couldn't grasp how that internal "shamanic" shift to peace and wisdom had happened to me or to any of the others in the room whose state of consciousness had been altered. The answer was a long time in coming; it wasn't until I learned about the most elusive cytoskeleton decades later that I began to understand.

The Real Deal—Science Meets Spiritual Practice

When we drum, laugh, move our bodies, make love, or experience any other form of pleasure, our bodies surge with chemicals called endorphins.[12] Molecules ebb and flow, brain waves change, cell tensions ease, and we re-create our emotional condition from the inside out. Our cells have the ability to use physical rhythm and the vibrating energy of our senses to create these pleasurable states. Energy shifts. Our muscles soften. We may even reach another level of being or knowing. Within our cellular scaffolding is where humming, drumming, light, movement, "vibes," and thoughts shift mind, body, and spirit.

Shape-Shifting

The underlying matrix of the cytoskeleton has roles beyond gene regulation. It is the "shape changer" and energy transformer; some say it's the seat of consciousness.

Shape change transmits information.

Shape-shift is an unusual word typically assigned to the shaman, magician, or mystic.[13] You may have read tales of Carlos Castaneda's Don Juan shape-shifting into a coyote or Harry Potter's godfather, Sirius, turning into a big black dog whenever the spirit moved him. While the idea is convenient in storytelling and myth, for our purposes *shape-shifting* means shifting our point of view and our emotional, mental, physical, and spiritual energy. According to many sacred teachings, when we change our body we shift our consciousness (and the reverse is also true). Our potential changes. Here science supports an ancient claim: if we move our bodies in certain prescribed ways, we can alter how we feel and what we are able to do. Our cellular kingdom changes. *It shape-shifts us.*

Ancient mystics and shamans discovered that certain tension-changing body postures improved overall well-being by uniting mind, body, and universal energies. To shift the mind and soul for this purpose, many traditions use complicated patterns of movement: among them shamanic dance, tai chi, yoga, and the ancient ritual postures explored by anthropologist Felicitas Goodman.

Carlos Castaneda, who purportedly spent years in Mexico studying sorcery and magic with Yaqui shaman Don Juan Matus, claimed that ancient physical practices can enable us to sense energy flows and shape-shift our bodies. By doing what he called *magical passes,* we can tune in to both inner and outer energies and effect a change of consciousness. Though he never talked about the *cellular* tensegrity we have discussed here, he nevertheless codified these techniques as tensegrity movements—this was the subject of the *Yoga Journal* article I discovered in the bookstore that momentous day. The term is appropriate for magical passes and the other body practices previously mentioned, all of which emphasize tensing, stretching, and relaxing muscles and organs—similar to the very movements our microscopic cells naturally undergo.

An obvious question arises: if we engage in magical passes, qigong, or dance, do we change our cells' tension, memory patterns, or genes? Do we alter their intelligence or their future? By moving our bodies, can we adjust our state of mind, energy field, or consciousness? Consider that the property of tensegrity within our cells may provide a new explanation for why staying physically active prolongs life; it alters our mood and energy. It expands our potential for pleasure, well-being, and peace. Movement can change our lives; Sri Aurobindo would call this "yoga for our cells."

Our cells change shape, move, grow, and "choose" what to do with the help of tensegrity. They manage us through tensing and releasing tension. Our bodies need to stretch and move to keep our tissues healthy and flexible. Yoga and other forms of movement as well as massage and chiropractic may be viewed as therapeutic interventions based on this principle. They bring us into our bodies, help us sort out thoughts, enhance energy, and encourage letting go of patterns that no longer serve us by tapping into what I call our *cellular shaman.* When we pay attention to the matrix of forces occurring inside and outside of us, we can change our lives. Here is where science and sacred wisdom meet.

Consider that one stiff muscle can change the structure of the whole. If you have ever pulled a calf muscle or stood up from a session at your desk with a pain in your neck, you know this is the case. Tense structures share a critical factor: their tension is continuously transmitted across all structures. The whole body reacts to a pain or pulled muscle. Muscles—which are bundles of cells—are able to shorten, lengthen, or freeze in place. A muscle held taut in one position reduces circulation of blood, breath, and information elsewhere in the body and can cause chronic pain. Rigid lung cells can hinder breathing.

> *These linings, wrappings, cables, and moorings are a continuous substance. Every single part of the body is connected to every other part by virtue of this network; every part of us is in its embrace.*
>
> —DEANE JUHAN *Job's Body*

◇◇

REFLECTION

By understanding the mysteries within our cellular matrix, we uncover a teaching message from our cells: stretch and move for renewal and change. How have you stretched yourself today?

◇◇

Tuning Our Strings

Our cells possess "tone" just as muscles do. To use an analogy, a stretched violin string produces different sounds when pressure is applied at different points along the string. In a similar fashion, a cell processes chemical signals differently depending on how much and where its cell strings are distorted or pressed.

Recalling the physical nature of our cellular matrix, we recognize that we are made of strings. Our cellular fabric shape-shifts us when we engage in physical, energetic, or shamanic transformative practices. *The strings of the universe can now include the strings of our cells.*

Only with personal experience can you find out if this is true.

Strings vibrate. Pluck a guitar string, and the adjacent strings will vibrate; the strings resonate with one another. The same is true of drums—strike one, and another close to it will vibrate in response. Consider that our cellular strings respond to movement, sound,

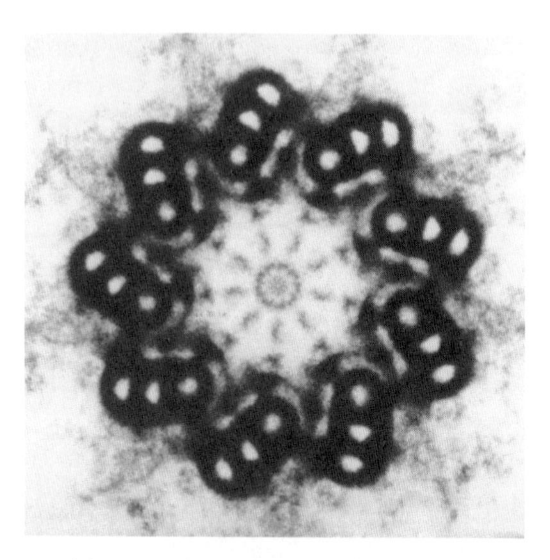

Figure 4.5 Two centrioles; notice the regular structure of nine sets of three microtubules; image by Don W. Fawcett/Photo Researchers, Inc.

humming, music, and chanting. Coming into harmony with our cells takes on a whole new meaning when you remember the resonating intelligence inside. Perhaps the scaffolding of our cells is the location where energy, movement, and vibrational healing take place.

Exploration

Strut Your Stuff

Hum, move, dance, get a massage. Discover what moves you to let go.

"Seeing" Energy

According to noted Northwestern University scientist Guenter Albrecht-Buehler, cellular movement is part of the cell's intelligence.[14] Cells seem to move intentionally toward each other: through a microscope you can see them touch and then slither away. To guide their movements, cells see and "read" each other's energy with bizarre "eyes," strange constructions of microtubules called centrioles (each cell has two). The centrioles were once believed to only guide cell division, but now it is thought that the centriole may also be the director of all cell movement. A unique mathematical construction, each centriole is built in series of threes (3^3)—twenty-seven tubes (nine triplets) arranged to form a hollow channel in the center (see figure 4.5).

Resembling twisted pipes, the centrioles' construction is certainly unusual; their function is even more so. It is said that our centrioles can detect infrared energy generated from neighboring cells, and this is what enables them to "see" each other energetically.[15] Through their centrioles' eyes, cells pick up heat and each other.

Bending Consciousness

Energy-sensing centriole pipes cross and twist, bend and flex. Cells nearby respond and do the same. Nobel Prize–winner Francis Crick and world-renowned physicist Sir Roger Penrose join Albrecht-Buehler in proposing that the centrioles transmit information by changing their shape as a result of electrons flowing from one end of the centriole tube to the other.[16] According to these scientists, the electron flow down our cell tubes is "consciousness."

Consciousness? What is it? Where is it? One idea about which there is basic agreement is that when we are awake and conscious, we have a consciousness. Following from this, Penrose and anesthesiologist Stuart Hameroff explored the theory that microtubules are engaged in human consciousness.[17] Hameroff provided cellular evidence by exploring the effects of anesthesia—ether and halothane—which freezes the microtubules in brain cells and induces sleep. The conscious aware state halts while survival-based brain functions remain active. So when microtubules in the brain are "frozen" by anesthetics, human alert consciousness disappears. The exciting role of microtubules and cellular fabric in consciousness is certainly a fertile area for research.

Clearing Our Pipes: The Shamanic Roto-Rooter

During my now many years of training as a "wannabe shaman," and even before I started down that path, depression has been my periodic companion, visiting at times when I wanted to change but couldn't—when I was stuck. Whenever I was rooted in the quagmire of old, unproductive habits and behaviors, my shaman-teacher Tomas would tell me, "Clear your pipes! Spirit can't move through you to initiate change until you clear your pipes."

I had no idea what pipes he was talking about or where they were, and I never asked. The truth was, I was embarrassed that I didn't know. Were the pipes my arteries and veins, my windpipe, my energy channels? I didn't think much about these enigmatic pipes; I wrote them off as yet another expression the shaman used that I didn't understand.

Regardless, over the years I learned fairly dependable strategies for "unsticking" my mind or mood in the darkest of times, *if* I employed them frequently enough. Chanting and shamanic dance have been pretty consistent change makers, continuing into the present. On a recent morning spent working on this manuscript, for example, I felt the old "stuck-ness" return. My ideas felt old, my body and mind stagnant. I took a break, lit a candle, burned some sage, and closed my eyes. I began to chant, feeling my chest, ribcage, and heart vibrate and hum. At the same moment, a thought percolated to the surface: my cells had to be vibrating too! The invisible, flexing, fiber-optic tubes and webbing within my cell sanctuaries must be bouncing away, smoothing out the kinks, flexing, their electrons streaming.

An "aha!" moment—the light goes on. These centrioles could be the pipes in the shaman's directive—clear your pipes! Chanting and humming activate your cellular shaman.

Calling Your Cellular Shaman

A cellular shaman pulls on the strings of the invisible,
and when those luminous filament are pulled
or pushed, everything changes.

A cellular shaman moves through, as archeologist, digging out old patterns and examining the remains—

To learn from, enjoy, prevent, ride through, or avoid altogether

Help your shaman choose.

Consider this: You have a particular pattern of behavior, such as reacting angrily to your partner or one of your parents, eating when you get tense, or chewing your fingernails while waiting in traffic. It's as if you have an attachment point that keeps bringing you back to the same old place again and again, all best intentions aside. Repetitive behaviors like these imprint on your body's cells.

Suppose you could cut loose from the broken record and create a new pattern to change your feelings, actions, and habitual responses. Psychotherapy can help—and so can engaging the cellular shaman.

Exploration

Drum a New Rhythm

Listen to a shamanic drumming tape, join a circle, and discover something new.

Body Prayer

Giving Thanks

Flex your wrists. Bend them back and forth.

Reach up toward the heavens; flex your wrists again. Give thanks.

Reach down to the earth, touching her, giving thanks.

Repeat this three times.

Each time, you can offer a different prayer of thanks or of intent: to feel grateful, to free yourself from pain, to take on a new discipline, to forgive. You'll know which prayer by listening to your shamanic wisdom.

Exploration

A Visualization Journey

To prepare for this journey, take some time to find a good place to sit or walk for a while. If you're walking, listen to where your feet want to guide you. Make yourself comfortable. Look around. Listen to the breeze. Feel the air or the sun on your skin. Take in the smells around you. Breathe in and out, letting the streams of air float easily together. Breathing in the universe unites us to our heritage, our home.

You can record the following instructions or simply recall them. Enjoy the journey and free your imagination.

Picture yourself entering a perfectly round structure, hide-covered and warm. There is a fire in the center, drums are softly beating, and the smell of piñon fills the air. You are home. You and your cells are peacefully at rest. As your eyes become accustomed to the dark, look around—people sit in a circle, drumming and chanting. Take a seat in the circle and notice who is in it with you. Drums beat the rhythm of your heart, and you breathe into your heart; it is resonating with the drumbeat.

Your heart cells beat in rhythm with the drum. Cells strum their tune throughout your entire body. You are breathing, moving, dancing, humming, and drumming your cells into a shared rhythm. Your mind is at peace in this cellular dance. You expand, connecting to a greater energy. Touch your own divinity and experience your own cellular shaman at play.

Enjoy this softened space and ask your cells, your shaman, and the people in the circle with you: "Is there anything I need to know or do to help expand my life?" Relax and listen. Receive whatever comes to you. Those sitting in the circle with you may have messages to guide your journey. Take in the peace and thank your cells for all they do. Thank the wisdom guides in your circle. When you are ready, shake out your hands and feet and come back to being aware and alive in the present moment. Then put what you learned into physical action to anchor it.

Reflection

When we move our bodies we begin creating new patterns; our cellular threads and neuromuscular wiring weave and anchor a new experience. We break the strands that hold us to old habits. Next time you find yourself reacting in the same old unwanted way, break the attachment to that behavior through sound or movement. The key is to begin!

What do I need to let go of?

Where is there too much tension in my life?

Am I attached to ideas, people, or habits that hinder my maturation?

What will expand me?

What contracts me?

Where am I being too repetitive?

Into the fabric of our cells our experiences
are woven and rewoven.

We change the weave and pattern depending on what
attracts and holds our attention and intention.

We are always in a state of creation.

What do you choose?

Untie the knots!

Break old strings and attachment points!

Weave new patterns!

Stretching Our Limits

I am visiting a friend in Santa Fe who suggests we go to an ancient sacred ritual site where she has done many ceremonies. What a wonderful idea! The day is cold yet sunny, the sky bright with blue winter light.

We arrive at one of the rare unguarded gates to Bandelier National Monument. Only a few cars are parked there, and we soon realize why; all the national parks and monuments are closed. The federal government has run out of money and shut them down to save a few dollars.

It's closed? How can this be? We have traveled all this way and want to pray there. I especially want to take photographs. I have snapped pictures of petroglyphs and pictographs in California and Arizona, and now is my chance to see what might await me at a sacred site in New Mexico. I have come to love investigating and documenting what our ancestors have left behind for us to learn from. In fact, I have been attempting to uncover a hidden language common to all cultures, human and otherwise.

My friend and I look at each other, then at the fence keeping us from our mission. The choice is obvious: climb over it! The people who belong to the few cars we see must have already done so.

We stroll up the entry path and begin a gentle climb upward. Then, after scampering over huge granite boulders, we come to a fork in the road—and a choice. Go up a ladder that has come into view, or head straight on the path and come down the ladder on our return. My fear of heights decides for me; if I go down it, I'll have to *look* down. The very thought petrifies me.

Up we climb. When we get to the top and I feel the solid mesa beneath my feet, we hear a voice. "Come back! The park is closed! Bandelier is closed!"

It's the park ranger, demanding that we leave the monument—*now!*

I argue with him for a few minutes, putting off the inevitable, dreaded trip down the ladder. "We pay taxes," I tell him. "This is our park, and we can visit it anytime. And we're not doing any harm."

"The park is closed! Come down now!"

Suddenly, for a moment, I step outside my fear and am swept with a feeling of tremendous loss. I feel the human pang of grief that comes from being exiled from sacred land. When I touch the earth at sacred sites I can sometimes sense the centuries-old spirit of life there, echoes from all the years of living, breeding, praying, playing, holding hands, loving. When I stand on sacred ground, I can remember. To be separated from it breaks my heart.

My arguments have not worked, though, and we must descend. Terrified, I begin to follow my friend down the ladder. I feel like I'm undertaking a sacred journey, a quest—and in fact, I am. Halfway down the ladder, I reach into my backpack with trembling hands, pull out my camera, and document my journey.

Most importantly, though, and the reason I mention it in this chapter, is that the moment reminds me we have the power to *stretch. We all do.* Fear does not have to paralyze me; my cells and I can choose to *move.* This power is within you and your cells too.

We can all make the choice to recall what we already know deep in our souls: we have come here to do it right, to ignite the heart of what is sacred and divine in each of us. Barriers of mind, culture, dogma, or even karma may prevent us from remembering, but the push-pull of our cells, the intelligence that dwells there, does not forget.

Chapter 5

Energy—Sustain

There is vitality, a life force, energy, a quickening that is translated through you into action and because there is only one of you in all of time this expression is unique. And if you block it, it will never exist through any other medium and it will be lost. It is not your business to determine how good it is, nor how valuable, nor how it compares with other expressions. It is your business to keep it yours, clearly and directly, to keep the channel open.

—MARTHA GRAHAM

So far, our scientific focus has been on the physical nature of our cells. We've learned that these mobile message carriers chat endlessly among their trillions of fellow cells. We've seen how they speak through molecules and movement, by changing shape, and even through vibration. But what about energy? In this chapter, we will traverse a variety of energetic terrains: the molecular energy of our cells, the bigger picture of vital energy and life force, and our personal relationship with energy—how we invest and renew our resources.

First let's consider how the view of our universe changed when Albert Einstein gave us the world's most famous equation, $E = mc^2$, where E = energy, m = mass (matter), and c = the speed of light measured in meters per second.[1]

What this famous equation tells us is that energy and matter (mass) are essentially interchangeable, different forms of the same thing. Matter becomes equal to energy only when it moves very, very fast. If matter could move fast enough, it could transform into energy—perhaps light energy. The equation also says that matter is packed full of energy.

For our trillions of cells to maintain themselves—and us—adequate molecules of energy must be available to them at all times. Where does this energy come from, and how do we manage it?

Defining Energy—More than Molecules

Scientists define energy as the ability to do work. It can be measured as heat, calories, joules, basal metabolic rate, and, in the cell, molecules of adenosine triphosphate. Physical energy has many forms: biological, chemical, thermal, electrical, nuclear, magnetic, and even quantum.

Energy, as we know, also encompasses the emotional realms—the energy of happiness, anger, passion, sadness, boredom, and enthusiasm. Energy also lies at the heart of the great Mysteries: the immeasurable life force, qi, or *prana;* love, soul, faith, and prayer. No matter what we call them, positive qualities of energy contribute to our well-being and joy for life. Diminished energy contributes to fatigue, depression, and low vitality.

As a biochemistry student, my first introduction to the concept of energy was learning how cells produce chemical energy through a very complicated sequence of events involving the breakdown of sugar into carbon dioxide and water. This process, taking place continuously, maintains the health and repair of our cellular sanctuaries.

Years after I gained an understanding about our cells' molecular energy production, I began teaching people stress management. It was then that I gained a deeper understanding that energy pervades who we are and what we do, mind and body, in microcosm and in macrocosm, from cell to soul. We are "energy beings" who need to know how to manage our personal energy and our global resources. When we develop awareness of the energetic forces within and around us, we can discover what drains us and what sustains us. And this gives us ample choice in how to wisely invest our own energy.

Some of my most personally meaningful explorations into energy came about when I was ill and seeking to understand the nature of healing. Almost every healing tradition I studied in my quest to get better came down to energy at its core: Chinese medicine, Reiki, hands-on healing, yoga, massage, chiropractic, indigenous practices, and more. In all of these healing practices, it was not molecular energy that was discussed; rather, it was an invisible force flowing through and around our bodies. This energetic force goes by many different names—*qi, kundalini, prana, kupuri,* and *num,* to name a few—yet all describe the same invisible power.

Qi (or *chi*) is the term used by traditional Chinese medicine (TCM) to describe the energy flowing through our bodies and everything else.[2] According to TCM, in the human body energy runs in channels called meridians, and disease occurs when energy stagnates or is thrown out of balance. Acupuncture is one way qi energy can be balanced; another is the practice of qigong. When I was first exposed to the notion of qi, it was just one more elusive concept to me, an imaginary quality whose existence I doubted—until I studied with a qigong master. After a few practice sessions, qi was no longer an obscure notion: I could *feel* it!

Qigong means "cultivating qi energy." It is a basic component of TCM, and in China where it originated, there are actually qigong doctors who specialize in teaching their students—they do not call them patients—these healing practices. Said to have originated in ancient shamanic dance, qigong also provided the foundation for tai chi. Its practice can reward us with a tangible experience of qi and the improved health that results. I attribute the fact that I have not had the flu in more than a decade to my daily practice of qigong.

According to TCM, the areas of the body where qi is closest to the surface are called gates; on the palm is the *laogong* gate or point. This is the principal point for emitting qi. It was here in the palms of my hands where I first felt a force that was different from my pulse, my breathing, or my beating heart. From our hands or the hands of others may be our first experience of the elusive invisible energy. And now I would like you to be able to have this experience. By the way, this is an especially good practice for skeptics!

Exploration

Take a few minutes to discover qi energy for yourself.

Sit or stand in a relaxed position.

Touch your fingers into the palms of your hand. Where your middle finger touches the center of your palm is the laogong point. Touch the energy centers of your hands and then relax your hand.

Now bring the palms of your hands together and rub briskly to stimulate warmth, circulation, and qi.

Next, extend your arms out in front of your chest, shoulder distance apart. One palm faces up, the other faces down. Elbows are slightly bent. Shoulders, arms, and hands are relaxed. Remember to breathe.

Open and close your hands about twenty times, like you are gently pumping them. With your arms still extended in front of you, reverse the direction of your palms so that the one facing up now faces down. Once more squeeze your hands twenty times. When you are finished, keep your hands closed.

Lower your arms, bringing your elbows close to your waist. Your closed hands face toward each other. Now, with palms facing, slowly open your hands. Gradually bring them closer together until you notice a sensation, a tingling, or a feeling of density between your hands. What do you feel?

Slowly move your hands away from each other until you no longer feel any sensation between them. Move them back and forth as if playing with an invisible ball.

This is qi.

People describe their experience of qi in a variety of ways: as heat, a pressure or force, tingling, "magnetism," pulsation, or warmth. You may feel nothing at all, and this could mean that you were tense during the exploration. Try it again after you've done some exercise or when you feel more relaxed. Even if you don't feel anything now, once you become aware of this quality in your hands, you may begin to notice an additional sensation or force from your palms when you exercise vigorously or practice the gentle movements of tai chi.

Skeptics say that qi and the other esoteric, immeasurable forms of energy do not exist, and healing modalities that employ them are nothing more than quackery. I always counter this assertion with one irrefutable scientific reality: until we had the technology to be able to *see* viruses, they, too, were hypothetical constructs. It took the development of the highly powerful electron microscope to prove that they existed. Until then, some "undetectable force" or "germ" was responsible for many diseases. Remember, science depends on objective measurements to declare something real. Once we are able to measure qi—which I believe one day we will learn how to do—perhaps the skeptics will be as convinced of its reality as they now are of the existence of invisible viruses.

In the meantime, if you are willing to continue to engage in direct experience as you did in the preceding exercise, you can receive the benefits of this invisible force through practices such as qigong and tai chi. You can also learn more about the growing body of Western scientific evidence of the effects of qi by visiting the National Center for Complementary and Alternative Medicine's website: nccam.nih.gov. One of its first studies showed that the acupuncture treatments used to relieve pain increased the body's endorphins: the pain-relieving molecules we touched on in the last chapter.

Universal Energy

All living things require energy to survive. And going back to $E = mc^2$, the first law of thermodynamics follows that energy cannot be created nor destroyed; it can only be transformed from one form to another.

Plants do a far better job at energy transformation than we do, converting solar energy into molecular energy. They use this energy to change simple chemicals in their environment—nitrogen, water, and carbon dioxide—into complex organic molecules needed to sustain life. This process is called photosynthesis. Plants are self-sustaining solar collectors that also provide nourishment for others.

We humans can use solar energy in only limited ways—to help our cells make vitamin D, perhaps to get a tan, and to improve our mood. Neither we nor other animals can use solar energy to convert simple molecules into food; those of us who walk, fly, and slither are wholly dependent on plants to do this for us. This is a tangible reminder of the interdependence of life on this planet and of how important it is for us to sustain the rainforests, farms, and other areas that burst with green energy. Our lives depend on it.

Our Cellular Energy—Unusual Origins

Today as we look at cells from the standpoint of molecular energy, our cytonaut self steps inside the cell and moves past its membrane and receptors. Secured within the cell matrix we see strange objects that, to my eye at least, look like flying sausages or perhaps creatures from outer space. These are the *mitochondria,* the energy generators of the cell (see figure 5.1). Every cell in our bodies except red blood cells contains mitochondria. Startling discoveries have indicated that these unique cellular "power plants" have different origins than the rest of the cell.

In fact, the mitochondria that are now our cells' powerhouses didn't start out that way.[3] Billions of years ago they were a kind of early bacteria, with the ability to convert the toxic oxygen in the environment of that

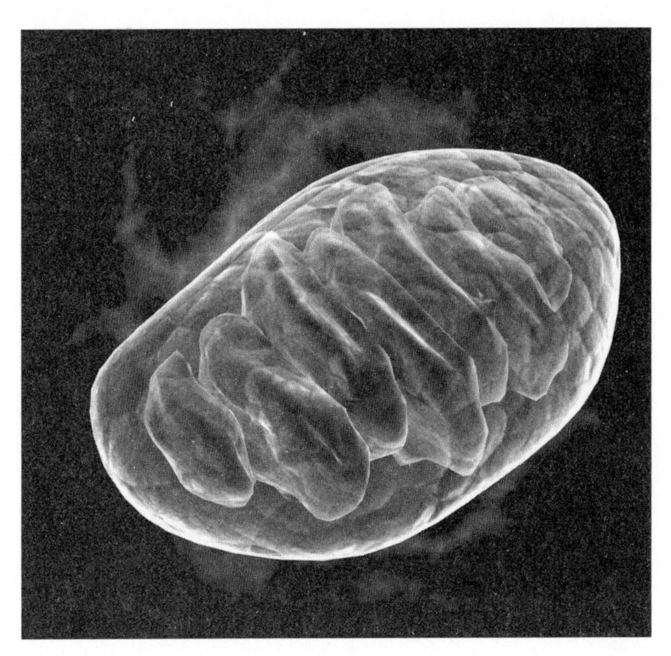

Figure 5.1 Mitochondrion

time into something useful. Then these "pre-bacteria" merged with early versions of living cells. The result of this new cohabitating scheme was that ancient organisms thrived, since each brought something new to the table of life; together they could do what neither could do alone, and they ultimately evolved into the cells we know today. The job of the mitochondria is to make the fuel our cells use to power everything they do. Once our cellular ancestors acquired mitochondria, they were able to produce so much fuel that they could get very big, much larger than bacteria.

Evidence for mitochondria's remarkable origins includes:

- Mitochondria contain their own genes and DNA, different from those found in the cell's nucleus.
- Their genes provide information dedicated solely to making energy.
- The membrane coating of the mitochondria contains unique lipid molecules found only in bacteria.

The nucleus is the primary and only other part of the cell that contains DNA—and its DNA differs significantly from the DNA in mitochondria. First of all, mitochondrial DNA (mtDNA) has a different shape; it's circular—ring shaped—rather than the spiral form it takes in the nucleus. (Bacterial DNA is also typically circular.) The only genetic information mtDNA contains is for producing energy and making more mitochondria. Even more unusual, we inherit mitochondrial DNA only from our mothers. Why doesn't Dad's mtDNA get passed on? Because the male's mitochondria are housed in the tail of the sperm, which doesn't enter the egg during fertilization. Only the egg holds keys to our energy production: it is Mom who lights the spark of molecular energetic life by passing along these round DNA molecules. We might think of her as the "Lady of the Rings." And we can thank those early creatures for being willing to form a collective, providing us with the means and energy to survive in our oxygen-rich environment.

Energy Production

Because of their unique origins and talents, mitochondria make it possible for our cells to transform the food we eat into high-energy fuel. All the work of our cells—to reproduce themselves; manufacture new parts and materials; move to confront a predator; transport molecules into and out of the cell; and keep our hearts beating, our eyes seeing, and our muscles contracting—requires energy. Every day, an astronomical number of mitochondria provide each of us with at least three pounds of molecular energy. About 1,000 molecules of ATP (see figure 5.2 and plate 10) are used every second, which means more than 15,000 grams every hour (technical note: 1,000 grams = 1 kilogram = 2.2 pounds). Since our cells contain only about 3 ounces of stored ATP—enough to power a ten-second sprint, they have a very dynamic recycling system that produces millions of molecules of ATP each hour. Some scientists say that we actually make our weight in ATP every day. Of course, if we are running a marathon, our cells work even harder to keep us "powered up." In fact, it's said that we have three times as many mitochondria as cells in our bodies. The cells that work the hardest, such as heart and muscle cells, require the most energy, and they house the most mitochondria—thousands of them in a single cell.

Our Energy Bank: ATP

Now we will get deeper into the chemistry of energy flow in our body's tiny sanctuaries. This gets pretty dense, so feel free to bypass this section if it requires too much of your mental energy. I continue to present the science of the cells in some detail for those who are curious to understand more about our cells' marvelous workings.

The energy that fuels our cells is stored in the chemical bonds of a high-energy molecule referred to earlier, called ATP or adenosine triphosphate (see plate 10 in the color insert). There are three phosphates attached to the "A" (adenosine) of ATP. The last two on the chain are held together by high-energy bonds (see figure 5.2). When one of these bonds breaks, energy is released to fuel the activities of the cell. It's not exactly clear what form of energy this is, yet this is where our cellular energy comes from.

The primary chemical source cells use to produce energy is sugar (glucose). Where the sugar comes from doesn't matter here; whether from honey, corn syrup, pasta, table sugar, a candy bar, or fruit, our cells must have glucose. Fats and proteins are secondary resources, and this is why when there's no sugar available to convert into energy, proteins or stored fat will be used for fuel. When people are starving, their cells begin to digest their muscle proteins, resulting in incredible weakness and risk of disease. Yet cells have "a sweet tooth" and prefer sugar. Through the process of oxidizing or "burning" sugar, energy is produced. Just as burning gasoline in a car's engine fuels the car to move, glucose, converted to usable energy, is the "gasoline" for our human vehicle.

For our cells to burn sugar and mitochondria to generate ATP, they require adequate nutritional intake of simple or complex carbohydrates

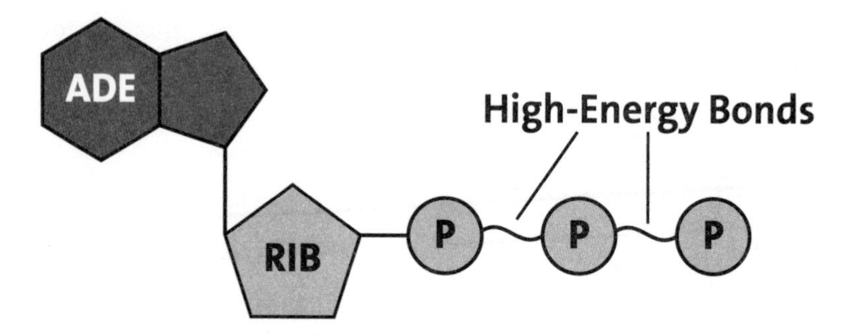

Figure 5.2 ATP and its high-energy bonds

(or protein or fats), water, oxygen, B vitamins, and coenzyme Q10 (also known as ubiquinone).

Basically, our cells have two ways of producing ATP: one in the absence of oxygen and one that requires oxygen. The very inefficient process that uses no oxygen is called anaerobic metabolism, or glycolysis. This occurs in the cytoplasm of the cell, not inside the mitochondria. Here, for every molecule of glucose, the cell produces two molecules of ATP. The glycolysis phase may also progress to the second, more efficient way of producing ATP that occurs in the mitochondria, the oxidative process called the Krebs cycle or oxidative phosphorylation. In the mitochondria, for every molecule of glucose, the cell now makes up to thirty-six molecules of ATP. What are the implications of this biochemical feature of our cells? When we are stressed, our cells take in less oxygen, and when that happens they can make only about one-tenth the amount of energy as when we are breathing deeply and relaxed.

Cells Teach Renewable Energy

Our cells, though not 100 percent efficient, attempt to reuse their resources again and again. Let's look at a molecular energy dance that involves changing energetic partners to see what I'm talking about.

To do their work, our cells generate ATP. What makes ATP useful is the release of energy held in its bonds. But after ATP's energy is released, it becomes ADP (adenosine diphosphate) which must be recycled back to ATP. Each cell is in a constant state of having to recycle "spent" energy, having used up about a billion molecules of ATP every few minutes. An intricate exchange takes place between ATP, ADP, and another high-energy molecule called creatine phosphate (see plate 11 in the color insert), which can be thought of as a kind of energy storage facility; our muscles and brain store creatine phosphate to have it readily available for quick bursts of energy. When ATP releases energy from one of its phosphate bonds, the energy is replenished when creatine phosphate provides a high-energy phosphate to reform ATP from ADP.

This chemical dance shows how resourceful the cell is. It wastes as little as possible and reuses whatever it can. Talk about environmentally aware and fuel efficient! Our cells are the original hybrids and recyclers.

ATP Production and Your "Real Life"

The magnitude of ATP production in our bodies is stunning. Our cells must generate about half our body weight in ATP every day. A cell uses about a million molecules of ATP that must be replenished every few minutes.

The biggest energy producers are our muscle cells, so they have the most mitochondria. A contracted, tense muscle cell produces only two molecules of ATP for every molecule of glucose "burned" because it's taking in very little oxygen, while a relaxed muscle in an aerobic state churns out about thirty molecules of ATP. If you are tired all the time, it could mean you're under a lot of stress; the cells become inefficient when making ATP anaerobically. Also, one of the by-products of such anaerobic metabolism is lactic acid, which builds up and accounts for why our muscles hurt after a lot of exercise and why we may feel anxious when our brain cells are saturated with it.

If you are under a lot of stress, what then? Take it as a sign that you need to take care of your cells. Take them for a brisk ten-minute walk, meditate, do deep breathing, or get a massage. It's said that a ten-minute walk can generate enough ATP for another ninety minutes. Relax and your cells will reward you with more energy as you bring them more oxygen. Soften your cells and remember that with relaxation, the tension of the cell's scaffolding eases, breathing deepens, and more energy is produced.

Large muscles are the primary source of energy; it may seem paradoxical, but the more you work them, the more energy you will have. At the same time, the more they work, the more mitochondria they will need. The paradox is solved when you consider that hardworking cells can produce more energy-producing mitochondria. Physical exercise relieves muscle tension and provides the cells with more oxygen. In essence, inputting energy to exercise, you are rewarded with more energy. In addition to giving you an infusion of energy, a regular exercise regimen can also reduce stress and help you achieve balance.

The Radical Nature of Energy Production

Energy production involves removing electrons from one molecule and passing them on, like a game of "hot potato." Sometimes during this game an electron escapes, and if not trapped it can damage the cell. Such unpaired electrons give rise to potentially dangerous free radicals, reactive molecules that can attack the structures of our cells in a process called *oxidative stress.* Wrinkles are one physical sign of oxidative stress. DNA mutations, macular degeneration of the eye, and even heart disease may be due in part to these oxidizing radical agents.

Mitochondria are one source of these reactive substances and are the first structures to be clobbered by them. Other cellular sources of free radicals are immune cells when they are hard at work attacking microorganisms. But besides being potentially dangerous, free radicals also serve as important signaling molecules for killing microorganisms.

In our cells' inherent wisdom, they are able to protect themselves from the danger of free radicals. They produce their own supply of protective *antioxidants:* molecules that can squelch or eliminate free radicals. Antioxidants that our cells make include lipoic acid, melatonin, superoxide dismutase, glutathione, and coenzyme Q10 (CoQ10). What our cells can't produce to protect from damaging free radicals, our food can provide, bringing other antioxidants to the rescue. Antioxidants in our food include vitamins A, C, and E, minerals such as selenium, and the polyphenolic pigments in red wine and deeply colored fruit like blueberries and prunes. In fact, the aging of wine is the result of oxidation, and the very substance that protects red wine from aging too quickly, resveratrol, may also protect cells in those who enjoy a glass of red wine.

In the partnership between plants and people, our food choices not only nourish us, they also protect our cells. One explanation for why people who eat a lot of fruits and vegetables tend to experience less chronic illness than those who do not is that they are consuming substances that protect against oxidative stress. In fact, a current theory for a cause of chronic illness is that free radicals cause inflammation, which in turn damages the cells.

Medical Molecules Mismanage Our Energy

One downside of modern medicine is that some common drugs have a side effect of suppressing the amount of ATP our cells can produce. This is especially a concern for people taking statin drugs, which inhibit the liver from making coenzyme Q10, an essential ingredient for ATP production.[4] Two side effects of statins are muscle pain and fatigue, most likely because of depleted levels of CoQ10.

Why is coenzyme Q10 so important? Because it has several roles in energy production—it is an essential ingredient in the electron transport process of ATP production, and it also acts as an antioxidant. If there is not enough CoQ10, less ATP is made. In addition, CoQ10 protects the mitochondria from damaging free radicals. One solution for the statin problem is to take a dietary supplement of CoQ10. Have a conversation with your physician if you are taking these medications. In clinical studies of people taking statins, a daily dose of 50 to 100 mg of CoQ10 helped remedy the problem of muscle pain and extreme fatigue.

Increased Requirements for Energy

Our bodies demand extra energy at certain times during our lives. Especially pay attention to what you eat and do, and how you replenish yourself, during these times:

- During growth and development of children, teens, and pregnant women
- When healing from wounds after surgical and dental procedures
- During chronic or acute illness
- When engaging in vigorous exercise
- When under stress

Definition

Stress: One definition is any situation that we perceive we don't have the resources to handle.

Stress and Tension Use Up Energy

You may be surprised to learn that scientists do not agree on what stress is.[5] (Then again, perhaps it's a given that scientists do not all agree on any definition.) Yet in terms of the physiology of stress, there is major acceptance that the hallmark of the fight-or-flight stress response is rapid mobilization of lifesaving energy. That means getting glucose into the bloodstream and getting tissues to produce more ATP. We need quick bursts of energy to get us out of a dangerous situation. Yet if that response lingers too long, leading to chronic stress, our energy stores will be depleted—and that compromises us on many levels.

Without enough energy, neither immune cells nor their potions can protect us against infection. If we have insufficient ATP in the brain, our mood and mental agility will suffer. If we are flooded by too many stress hormones, bones, heart, and gut all suffer. You get the idea: long-term stress simply drains us.

Stress causes physical tension and shallower breathing, and it gets the heart pumping more rapidly. Tension, be it physical or mental, is a major factor in unnecessary energy loss. As we have seen in this chapter, tense muscle cells use energy quickly and replenish it inefficiently. The very acts of thinking and coping consume energy, and when our energy is used up, we have greater difficulty solving problems and meeting challenges; we may simply lack the mental "juice" to handle the situation.

In teaching this concept to thousands of people, I have consistently seen that people instinctively know what their energy level is without any prompting or definitions of energy from me. You can answer the key question right now, just as they have: on a scale of 1 (*lowest*) to 10 (*highest*), how would you rate your energy?

When I began testing this question, I'd ask it first thing in the morning of a daylong seminar. Most people responded that their energy was pretty high. After lunch—that was another story. More than 75 percent of them reported very low energy, and the solution for getting them through the rest of the afternoon was to teach a short energy-enhancing qigong practice that you will have the opportunity to explore later in this chapter.

Energy Awareness Mapping

The energy graph shown in figure 5.3 is adapted from psychologist Robert Thayer's fascinating studies on the roles of energy and tension in coping with life.[6] People challenged by an unrelenting problem with no clear end in sight—going through divorce, losing a job, or putting a parent in a nursing home—were asked to graph their energy, mood, and tension

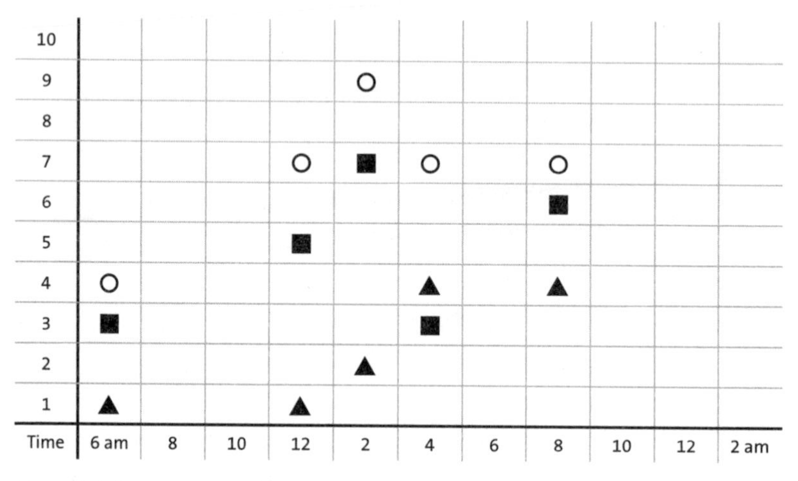

Figure 5.3 Energy graph with sample data: energy (squares), mood (circles) and tension (triangles) plotted throughout the day

throughout the day. They also indicated on the graph the severity of whatever problem they were grappling with. Universally, at high tension–low energy times of the day, in the subjects' minds problems loomed large and catastrophic. The problem itself had not changed; their perception of it had. When their energy was drained by tension, people felt overwhelmed, helpless, and pessimistic. In other words, the combination of high tension and low energy weakens our ability to cope with stress and life.

OK then, now we know. So what can we do about it? The first step is knowing when we are vulnerable. Energy mapping is an effective strategy for becoming more aware of your typical daily energy patterns.

Exploration

Mapping Your Energy

Keep an energy graph every day for a week, using the template provided in appendix 1 (page 217) to monitor your energy, tension, and mood. Make seven copies of this graph so you can follow your energy rhythms for at least a week.

Track these qualities at about the same time five or six times each day; times are indicated across the bottom of the graph.

Mark on the graph how you perceive your energy, tension, and mood using a scale of 1 (*lowest*) to 10 (*highest*). You might use one color to mark your energy and different colors or shapes for tension and mood.

Once you know your usual energy patterns, you can do another week of mapping to discover how practicing qigong, taking a walk, drinking coffee or alcohol, eating candy, or anything else you choose to do affects all three. Note whatever your added activity was on your energy graph. When you achieve greater awareness in this way, you add to your own storehouse of wisdom, and you can choose to make changes that better support your life.

I had a student who was in the habit of eating a donut to boost his energy as soon as he got to his part-time job. When he did energy mapping for a few days, he was surprised to see that he got an initial jolt of energy from the tempting, empty calories but that it dropped quickly afterward. A donut is a downer? Yes. And this convinced him to kick the habit. Sometimes all we need is a little "data" to show us what's going on and convince us to make a change. Keeping track supports us in knowing the impact of our choices, inside and out.

When I first did my own energy maps, I discovered that my peak energy occurred after 10:00 p.m. I had always known I was a night person, but seeing it in black and white allowed me to be easier on myself when I had a hard time getting started in the morning. One of my remedies for the morning sluggishness was doing qigong and the body prayers I offer in this book. Now my mornings blossom a bit more easily as I receive a burst of energy from practices that connect body and spirit.

Suppose you discover that your typical pattern is an energy crash with a lot of tension in late afternoon: the four o'clock blahs. Once you know this pattern, you can practice proactive stewardship by planning important events or meetings at other times. Or, if you must do something important at a low-energy time, you can manage your internal resources beforehand. To increase energy and lower your tension, do a relaxation practice or take a short, brisk walk. With some experimentation, you will soon learn what best renews and sustains you, and you will be rewarded with a greater sense of control and the ability to harness your inner resources. If you believe you have the ability to manage a challenging situation, this attitude itself can minimize your stress.

Becoming familiar with our own energy rhythms and knowing energy-enhancing practices can change our relationship to stress and daily challenges. Rather than being victims of surges of tension and troughs of energy, we learn when and how to regenerate energy, release stress, and relax our bodies and minds. It is important for all of us to know how to rest and renew ourselves and our cells.

Managing and Sustaining Energy

Understanding our energy rhythms is an important step in making healthy choices and living a more manageable and enjoyable life. There are three steps involved: knowing, being, and doing.

How to *know* (awareness, tuning in):

- Energy maps
- Body clues: Pulse, breath, hand temperature, muscle tension
- Mind clues: Who and what drains you? Who and what sustains you?

How to *be:*

- Exercise
- Meditation
- Stress release
- Qi
- Laughter
- Tai chi

In addition to the items above that will help you sustain your cells' precious energy, you can *do* any of the following.

How to *do* (energy renewal):

- Develop your own self-care plan
- Spend quiet time alone
- Do less, say no, *be*
- Learn mindfulness meditation or another meditation practice
- Reconnect with a spiritual source
- Recharge your batteries daily
- Walk
- Hold one focused, connected, and meaningful conversation each day
- Play, seek pleasure, and laugh
- Value and cherish yourself
- Ask yourself what gives your life joy and meaning; do something in pursuit of this each day
- Make a list of what brings you happiness, and at those fretful times when you might be drawn to a destructive pattern of behavior, choose a joyful activity instead

◇◇

Body Prayer

A Simple Qigong Practice for Cultivating Energy[7]

You will find complete instructions for this practice in appendix 2. (It is also available on my CD, *Molecular Messengers of the Heart*.) The following is a brief listing of each of the nine sequences in this series. Many can be done alone, or they can be done together as a complete series. The series includes:

The Basic Posture: Standing Home Alignment

Rooting and Spiraling: Waist Circles

Opening to Breath

Energy Wash (This part of the sequence is perfect to do when you want to relieve the mind of unwelcome thoughts or stress.)

Sipping Qi (This is like a reversal of the Energy Wash.)

Core Wave

Heart Thymus Wave

Integration: Balancing Yin and Yang, Right and Left Hemispheres (This balances the right and left hemispheres of the brain and is equivalent to alternate nostril breathing in yoga.)

Gathering and Storing the Qi: Closing the Circuits

REFLECTION

The following reflections guide you in taking what you have learned about how your cells create and replenish energy into a wider view of how you use energy.

How do I invest my energy?

How do I recycle resources in my home?

What sustains and restores me?

What or who wastes my energy?

How am I renewing my energy?

When I am spent, what replenishes me?

How do I want to invest my resources?

Energy and Energy

Most popularized tai chi, qigong, and yoga exercises are actually based on spiritual practices and the elusive forms of energy, not molecules or the workings of our cells. Because of the popularity of yoga and tai chi and the presumed benefits for lessening stress, Western medical researchers began examining their effects on physiologic measurements. It has been clearly shown that these practices enhance flexibility and aerobic capacity and improve our ability to relax. Elders who practice tai chi improve their balance and have fewer falls and hence a better quality of life.[8] So even though these energy practices were originally developed for spiritual deepening, once brought to the West, we demonstrated their body-mind and physiological effects. Our cells use all the energy that pervades their environment—qi and ATP. And we can too.

Fields of Energy

When we and our cells practice qigong with others, we generate a field of qi. Have you ever walked into a room and felt unpleasant "weird vibes" or absolutely at peace? Your experience is a reflection of both your sensitivity and the energy of the space. Walk into a church where incense is burning, candles are lit, and the essence of sanctuary surrounds you—the energy of hundreds of people who have prayed there. Enter a rowdy bar, and the energy screams at you. Many would call this "the field."

If you meditate, you well know what a different experience it is to meditate in a group compared to doing it alone. If you sing or chant, you know the difference between soloing and being part of a community of voices, surrounded by and becoming part of the sounds of many. Similarly, when we pray with others, we bring a level of focus that changes the experience for all. We might refer to the effect we create as a field.

Esteemed biologist Rupert Sheldrake speaks of a controversial field of energy that pervades the planet; he calls this the *morphogenetic field*.[9] One example he gives of the effects of this theoretical field is crystallization. When a newly synthesized molecule was made in a laboratory, it took months for it to crystallize. But then, as more labs around the world made this same chemical, the time span to crystallize shortened considerably. This didn't happen because one lab told another how to do it. Instead, as Rupert explains the phenomenon, the molecules themselves "learned" by means of the common field of energy.

When I first began teaching a particularly difficult qigong movement—the integration move—it took weeks for a class to get it. Then, the more large groups I taught across the country, the quicker people learned, until they got it the first time. I believe this accelerated learning was a result of the morphogenetic field; through repeated work, we had changed the field globally. Think of runner Roger Bannister, the first man to break the four-minute mile—considered an impossible feat at the time. Yet soon after he broke that barrier, many others were able to achieve it and then attain even faster running times. Was it because beliefs were shattered about what was possible, or had the change in the energetic field shaped new abilities or possibilities?

We can interpret this as meaning that what we learn or do on one part of the globe has an energetic or cosmic influence elsewhere. This is invisible energy, the Mystery, at work in ways we do not yet fully understand.

We each create a field of energy, regardless of how we name it. All living things generate an electromagnetic field around them. When I was first studying qigong, our teacher took us out to a grove of trees overlooking the Pacific Ocean. He told us to stand before a tree and use our hands to gather qi or sense energy from different parts of the tree. I was surprised to feel differences. He also put produce in a paper bag and had us feel the field of energy of the hidden mystery substance inside. Here was another lesson for this original skeptic: it was unmistakable—garlic and ginger had "hot" energy while an apple had "cool" energy. Was this the morphogenetic field or qi at work? I cannot say. But this kind of experience was proof to me that everything has an energy signature. When we are receptive, we may even notice that it's there.

When I pray in the redwoods, I feel a sense of peace that I don't feel in my house. The trees, the forest, the ocean, our gardens, and our pets all share energy with us. We only have to take the time to be with them to know and embrace this gift.

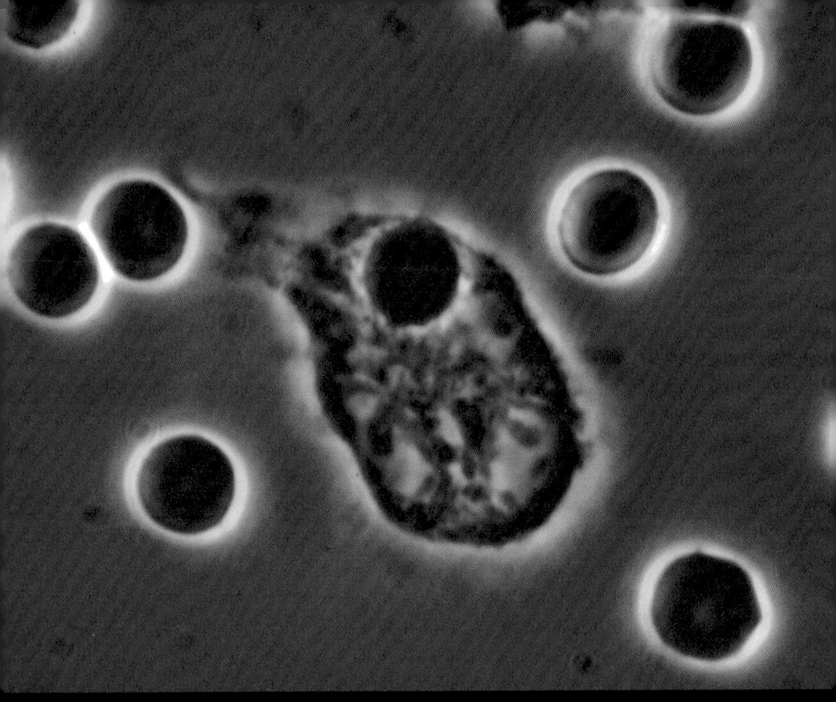

Plate 1 First cell photograph: Human white blood cell recognizing and discovering smaller cells from another species

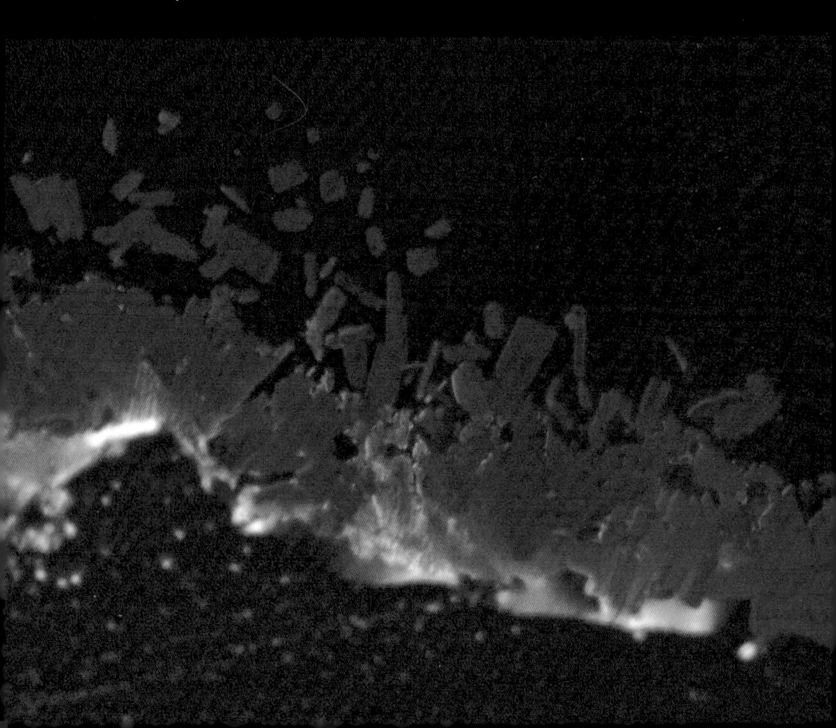

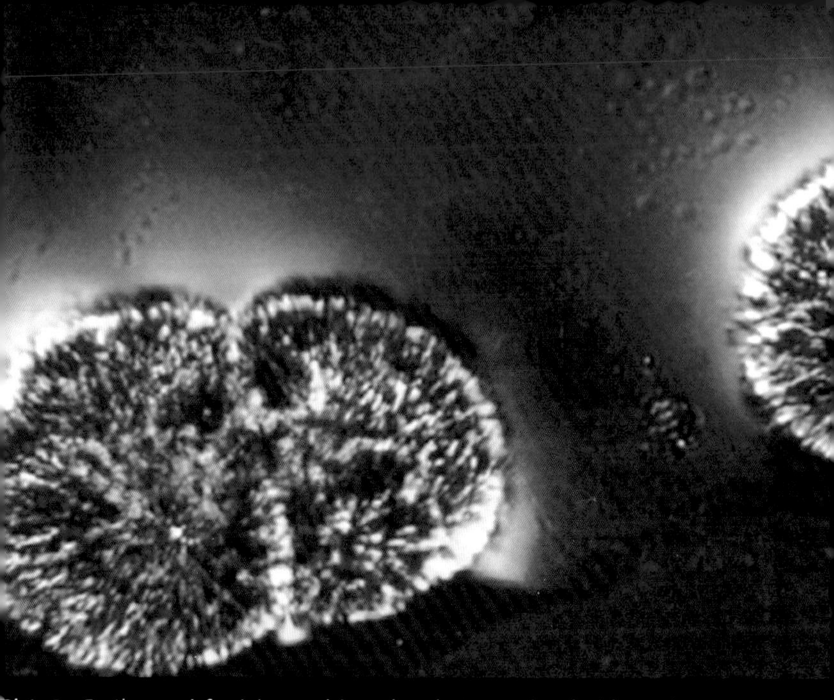

Plate 3a Earth, round, feminine—calcium phosphate associated with Capricorn

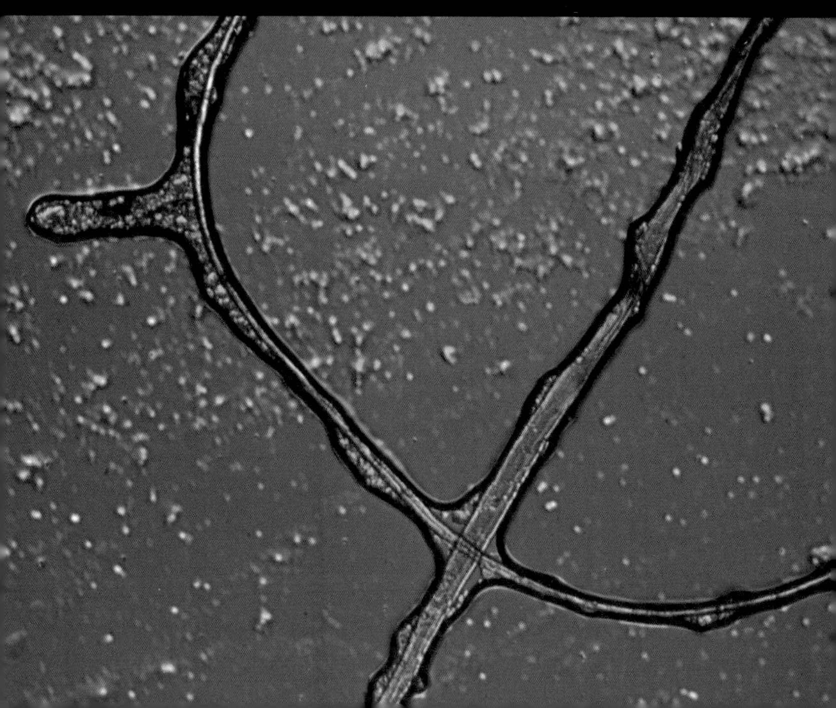

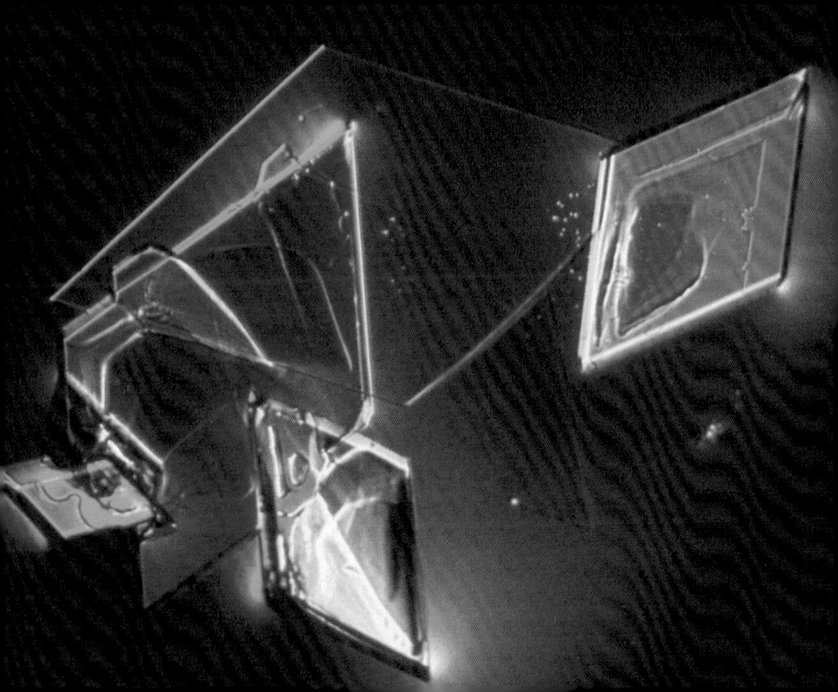

Plate 6 Sucrose, sweet taste

Plate 7 Malic acid, sour taste

Plate 8 Adrenaline

Plate 9 Caffeine, bitter taste

Plate 10 The molecule ATP—adenosine triphosphate

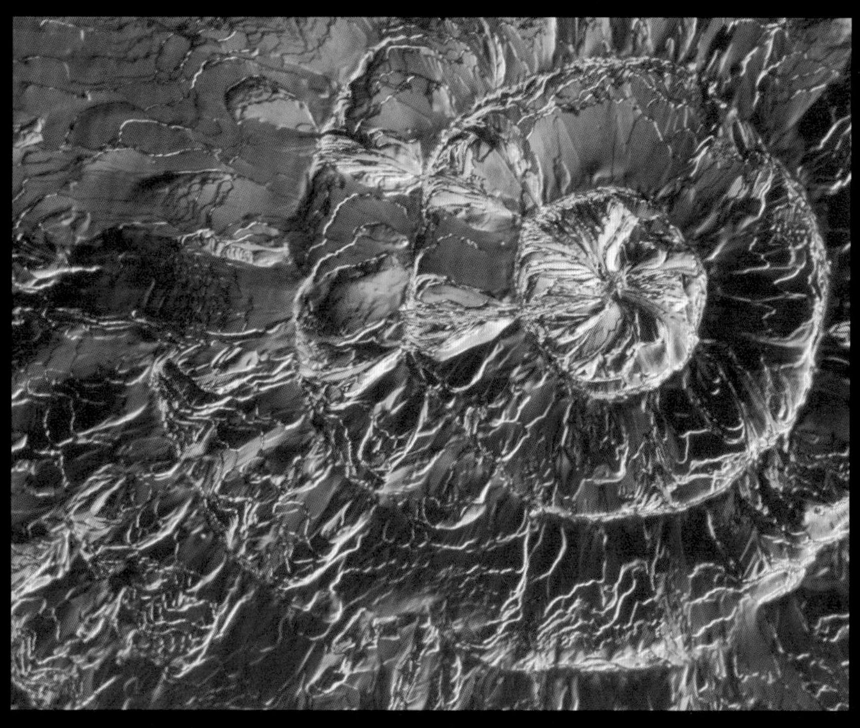

Plate 11 Creatine Phosphate—stores energy in our cells

Sacred Energy

We end this chapter where it began: asking about the nature of the energy of life. Where does it come from? Scientists argue about molecules and electromagnetic fields; theologians argue about soul and the God force. What is inarguable is that our trillions of cells are tiny cauldrons of life-giving energy until the day we die.

Perhaps a way to consider life's energy is to examine what happens at the moment of death. If you have witnessed the death of a loved one as I have, you will have indelible recollections of what death looked like, perhaps what it sounded like, and how you felt. And you will have impressions—maybe unexpected, and probably beyond articulation—of what has taken place. One minute your dear friend is alive, and the next moment he releases his life force, exhaling it away. The light that has shone through his eyes goes out. His body, his cells and molecules, remain, but his vitality—the sacred energy that has animated him—has traveled on. The instant of death is both a sad and sacred moment. Is this life force our soul?

In their quest to understand what it is that has left a person after the light retreats, people have gone to the extent of measuring the weight of an individual before and after death. They found no difference, of course. The infinitesimal spirit, the powerful energy of life, is weightless, and it is visible only in how it animates a person. My little friend Alvaro's sister had profound questions for me. "Where will he go when he dies?" she wanted to know. I couldn't even answer any such questions for myself: what *is* the energetic force that has animated his little body, and where *will* it go? Perhaps, looking again at $E = mc^2$, we can speculate that the life energy is transformed into light.

We may never find scientific answers to these questions, and that's OK. What we do have is the option, now, while we live, to view the thrumming in our cells as divine energy—the sacred spark that the mystic rabbis speak of as God's gift to each of us. The way we use and invest this energy while we are here on earth allows our spirits to live on: in the love we have given and shared and in the legacies we leave behind. The ways in which we have spent our brief journey here, loving and caring for others and our planet, ensure that traces of our divine energy will live on long after our physical bodies are gone.

In touching and animating our world, we return a host of favors. Dying stars give us our elemental substance; trees and plants gift us with breath and food; friends and family nurture our hearts and our spirits. In receiving these miracles without question, our sacred cells vibrate with energy and intelligence, and we are touched by the divine.

Energy is indeed present in all living things. Living organisms draw it from their environment. . . . They accumulate it in their own bodies and use it to power their movements and behavior. When they die, the energy . . . in their bodies is released to continue its way in other forms. The flow of energy . . . is part of the cosmic flux, and the energy within you will flow on after you are dead and gone, taking endless new forms.

—RUPERT SHELDRAKE *The Rebirth of Nature*

Chapter 6

Purpose–Create

The blueprints, detailed instructions, and job orders for building you from scratch would fill about 1,000 encyclopedia volumes if written in English. Yet every cell in your body has a set of these encyclopedias.

—CARL SAGAN *The Demon-Haunted World*

In this chapter, we will go deeper into the skills our cells possess as we discuss the role our genes play in cellular life. We will also discover that though our genes and DNA have become modern cultural icons, they may also mirror ancient metaphysical information still accessed today.

Before either the membrane or cytoskeleton was considered to be the key to cellular intelligence, our genes—a scripture of coded information—held that prominent position. But genes are passive; they can do nothing unless acted on by something else—molecules, the cytoskeleton, the environment, mind, or movement. Remember from chapter 4 that a cell's genetic expression determines whether the cell will reproduce or mature, manufacturing the markings and makings of a fully developed cell. Though a cell carries genetic abilities for both, it can only do one thing at a time: reproduce *or* mature. And of course, the final option,

Figure 6.1 Photomicrograph of DNA from calf thymus

death, is also written in the codes of our genes. Hidden in plain sight are the architectural clues in DNA that reveal secret codes and the mysteries of life and death (see figure 6.1).

I have been intrigued with DNA for decades from scientific, artistic, and metaphysical perspectives. As a photographer, my earliest "art" images through the microscope captured the beauty of spiraling molecules of DNA. As a research scientist, I explored whether we could change how genes express themselves within human cells. Having seen the devastating effects of chemotherapy and radiation in people with cancer, I was eager to test the possibility of another trail to treatment: rendering cancer cells normal. Instead of killing them—along with some healthy cells—could we program them differently? And could we do this with simple, benign natural substances?

There are basically two kinds of genes that we know of now: those that carry information about structure and those that regulate cell growth. In the case of cancer, I wanted to see if we could find a way to make the regulator genes switch programs. In other words, could cancer genes be turned off while healthy genes are turned back on? My colleagues told me to give up the idea of such "way-out holistic research." Yet I received major funding from the National Cancer Institute to test this approach. This was a very pleasant surprise, since my approach to investigating and treating cancer this way was not very common at the time, in the early 1980s.

My research showed that some human leukemia cells (malignant white blood cells) will indeed acquire traits of normal cells when treated with benign chemicals and hormones.[1] I was excited to learn from these discoveries that yes, we could change genetic expression in the test tube, yet I was acutely aware that much more extensive research was needed. But continuing to explore this line of research would entail using radioactivity, something I did not want to be involved with. Instead, I chose to leave basic scientific research and—to even my own surprise—began exploring whether ancient healing strategies labeled "unconventional" might help people with cancer. My goal at the time was to uncover the scientific underpinnings of ancient methods and apply these findings at the cellular level to the problem of cancer.

Then, as I began to work with *people* with cancer instead of their cells in the lab, my focus shifted yet again. I wanted to find ways to enhance a person's quality of life *whether or not* cancer cells changed their malignant course. How could we make life better even if disease persisted? Meanwhile, other scientists carried on their work at the cellular level, and as you travel through this chapter you will learn that they have indeed begun to effect genetic changes in malignant cells.

A Requirement for Life

You may recall that the requirements for life include the ability to reproduce. What facilitates that process is held in the nucleus of our cells, and like all cellular mechanisms, reproduction involves a coordinated effort that engages other parts of the cell. Yes, hidden in our genes, within the nucleus, are the instructions needed for cell duplication; yet for cells to reproduce, growth-regulating genes must first receive signals from the coordinated activities of the cell membrane and the cytoskeleton, which tell them, "It's time to express yourself!" Remember that genes are simply encoded instructions that require a "reader" of the code to act on them.

Architectural Design: The Language of Life

In the "heart" of each sacred cell—its nucleus—are the inherited instructions for making the entire organism, packaged into genes. DNA is the amazing molecule that encodes our genetic inheritance. About thirty years ago, scientists set out to decipher and catalog all the human genes through what is called the Human Genome Project.[2] (A genome contains the entire genetic code for an organism.) It is estimated that the human genome contains twenty-five to thirty thousand genes. The genome of a mouse has twenty-five thousand genes, a roundworm, nineteen thousand genes, and the single-celled *E. coli* bacteria's genome contains about five thousand genes.

Definitions

Cell nucleus: A membrane-enclosed region within the cell that is the sanctuary of genetic information.

Genome: Represents the entire genetic repertoire of a species.

Gene: The basic unit of genetic information.

Chromosomes: Genes and regulatory proteins are packaged in chromosomes. Human cells contain forty-six chromosomes.

DNA: The long, threadlike molecule that encodes the biological information of the genes.

The genetic code: Consists of a sequence of three-letter "words" written one after another along the DNA strand.

Genes are stretches of DNA that contain instructions for all the body's proteins, both structural and regulator proteins. Encoded by DNA, proteins are large, complex molecules made up of smaller subunits called amino acids. They perform most cellular activities and comprise the bulk of our cellular structures. Our hair is made of protein as are the enzymes in our saliva that break down those crackers we eat into useable molecules for our cells. Proteins are the regulators and messengers, identity markers and receptors. Without protein, life would not exist. The human body contains at least a hundred thousand different proteins.

Each gene reveals a different "packet" of encoded information necessary for your body to grow and work. Structural genes contain the recipes for how you look: the color of your eyes, how tall you are, the shape of your nose, the red protein hemoglobin, the keratin of your hair, the collagen in your skin, and more. Regulator genes tell the cells when to grow and when to stop growing. They also include the stop/start genes that indicate the beginning and end of a particular sequence of information. And yet, in the human genome, 98 percent of the genetic material has no apparent function; only 2 percent of our genes hold codes for all our proteins.[3] In the past, the non-coding genetic material was called "junk DNA"; now scientists are beginning to study what the rest of our genetic script does.

All cells in your body contain the same DNA and identical genetic information. Mature red blood cells are the exception: since they have no nucleus, they have no DNA. Only when immature red blood cells are developing in the bone marrow do they contain a nucleus.

Genes are packaged at specific locations in structures called *chromosomes* (see figure 6.2). You might think of genes as all the entries in the entire phonebook, while the chromosomes represent a single page. The genetic code represents the words on the page.

Chromosomes are long strings of genes. The cells within the human body contain forty-six chromosomes. Our chromosomes come in pairs, and so do our genes: half come from Dad and half from Mom. We actually have twenty-two pairs of what are called autosomal chromosomes (non-sex chromosomes) and a pair of sex chromosomes that determine

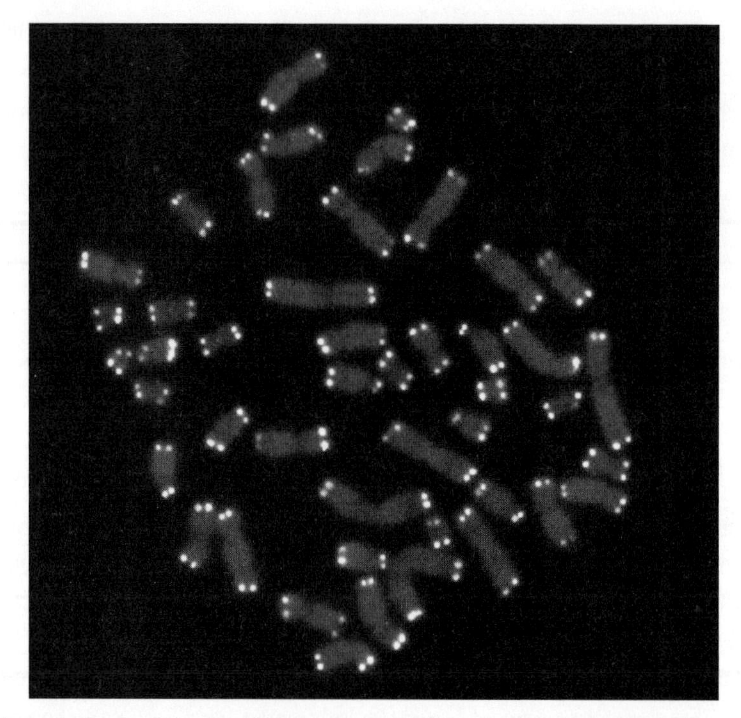

Figure 6.2 Forty-six human chromosomes with the telomeres appearing as bright dots at the ends of each chromosome; image by Hesed Padilla and Thomas Reid

our gender: the X and Y genes.[4] A female will have two X genes, while a male has an X and a Y. When a cell starts to divide, the chromosomes organize themselves along the center of the cell to be duplicated and shared between the new cells.

Metaphysical Secrets in Our Cells

Sacred Numbers

If you are intrigued by numerology, you may be surprised to learn that the number twenty-two is represented by the number of pairs of human autosomal chromosomes. In the divinatory tarot, twenty-two is considered a master number and is reflected in twenty-two cards of the major arcana, each of which represents a significant symbolic step along our life journey. There are also twenty-two letters in the Hebrew alphabet. You'll find more about sacred numbers in chapter 8.[5]

DNA and Its Divine Designs: Molecular Monogamy

Long strands of DNA are segmented into single genes that provide the blueprint for a specific protein.[6] Each DNA thread wraps around its complementary twin, creating DNA's distinctive double-stranded spiraling shape (see figure 6.3). To envision what DNA looks like, imagine two long threads coiled together in a double helix, like a twisted rope ladder with rigid rungs. One rope carries the code; the other strand complements it, acting as a kind of keeper and protector of the whole. Double-stranded molecules are much harder to destroy than a single strand. When a cell begins to reproduce, the DNA must be duplicated. To accomplish this, the two strands unwind and each serves as a template for a new double-stranded DNA.

The spiral is an efficient, compact way to safely store the astonishing amount of information the nucleus contains. The stretched-out DNA from one person is billions of miles long. How do we know that? Each human cell has around six feet of tightly coiled DNA, and if we estimate 10 trillion cells in an average person, that means that if we unwound all of that DNA and laid the strands end to end, we'd get around 60 trillion feet or about 11 billion miles of DNA. Your DNA could stretch from here to the sun and back at least sixty times! The Bible talks about Jacob's ladder stretching between earth and heaven. When Jacob dreamed of this ladder, was he seeing the spiral staircase of DNA? The ladder image holds great symbolic meaning for Jews, Christians, and Muslims, a connection to God and the spiritual path. And since we know one person's DNA could reach to the "heavens," perhaps this is another example of sacred secrets hidden within our cells.

Since DNA is a blueprint, its design must be precise. It is constructed from four molecular building blocks called nucleotide bases.

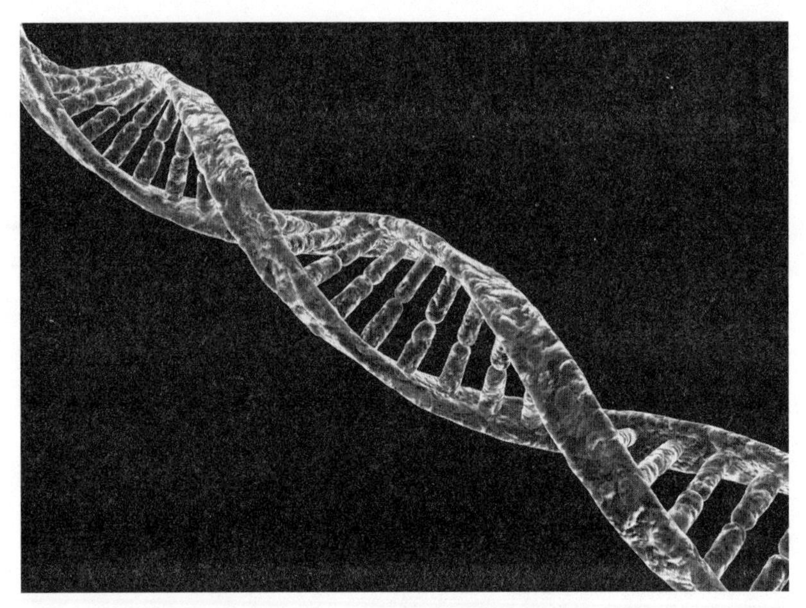

Figure 6.3 Spiraling DNA; the lines connecting the two strands represent the bonds between the nucleotide bases

The names of these bases are adenine (A), thymine (T), cytosine (C), and guanine (G). (The letters are used as shorthand to write out the genetic code.) The order of the bases along the strand of DNA carries the encoded information.

Here's another example of chemical complementarity: the building blocks are "monogamous"; each base can partner with only one other. Where there is a T on one strand, its partner on the other strand must be A; where G is on one side, its partner must be C. There are no exceptions to this exclusive union. Pairing with the wrong partner signals the need for separation and calls into play a potent repair system to separate the ill-suited partners. An advantage of such specific partner links is that it permits rapid detection of any mistake. With a partner error, there's a "kink" in the regular undulating spiral architecture of the DNA, and this triggers repair genes to come to the rescue. Molecular monogamy and the architectural helical design help ensure the integrity of our DNA.

The Code of the Winding Staircase

If you picture DNA as a winding ladder, you can see how one side of the staircase must perfectly match and complement the other. The sequence of the bases on one strand holds the genetic instructions. Its protective partner strand holds the complementary sequence.

Let's look at an imaginary genetic sequence: ATAGGCTTT. Its complementary partner would have to be TATCCGAAA (remember the monogamous relationship). The divine wisdom of nature rules how the two strands are linked.

> *The entire genetic library of this world . . . the structure of every living thing is reducible, ultimately to these four. . . . a very simple code but a very long message.*
>
> —CARL DJERASSI *The Bourbaki Gambit*

The same four bases that encrypt the genetic code—A, T, C, G—appear throughout the living kingdom.[7] Whether a bee or a hippo, an oak tree or a mosquito, everything on this planet uses the same encoding system; it's only the arrangement of the letters that distinguishes man from mouse. Though the computer holds its vast information in a binary code of 0s and 1s, the more complex DNA code is made of three units called triplets or codons. These encoded base letters signify which of the twenty amino acids is to be used in the building of proteins. There are sixty-four possible combinations of the three letters; all sixty-four are assigned either to amino acids or to start/stop signals. Imagine reading a hodgepodge of letters that encode specific information. You would have to know where to start reading the code and where to stop.

The sequence of these three-letter codons along the gene strand enables the cells to assemble amino acids in the correct order to construct strands of protein or a polypeptide (a small protein).

Here is the three-letter breakdown of a piece of DNA reading AAAATGCGTTCG.

Snippets of this code:	AAA	ATG	CGT	TCG
The complementary strand:	TTT	TAC	GCA	AGC
A complementary strand with errors:	GTT	TGG	GCA	AGA

The universal nature of the genetic code means that each species treats any new DNA as its own, generating millions of copies of these genes. In fact, one way viruses do their damage is to inject their genetic material into living cells and voilà—our own cellular machinery copies their genetic information and makes new viruses.

Metaphysical Secrets in Our Cells

Sacred Codes

Our cells reveal another cosmic "coincidence." Like the genetic code of triplets and sixty-four possible sequences, the ancient divinatory system of the *I Ching* depends on a coding system of trigrams: combinations of three different solid or broken lines (see figure 6.4). Trigrams, a group of three letters or lines, provide for sixty-four possible combinations. Both the genetic code and the I Ching present sixty-four combinations. Once more we have an ancient metaphysical system that could be interpreted as mirroring what takes place in our living cells. Both systems, cellular and metaphysical, reflect change.

Genetic Expression and Stem Cells

Every cell in your body has the same genes, and all have the same genetic potential. What distinguishes cells is which genes and proteins get expressed. Stem cells can use all the genetic information we possess and have the potential to become any cell type; the molecular environment these cells are placed in will determine whether the stem cells become blood cells or kidney cells, heart or bone. Once a cell becomes specialized or differentiated, however, it usually loses its ability to develop into other cell types. After the "purpose" of a cell is determined, it will express only features of that particular cell type. Immature red blood cells in the bone marrow can only develop into mature red blood cells. They can't revert back to stem cells and become white blood cells. Only immature red blood cells can make the protein hemoglobin. Dedicated skin cells, even though they have the necessary genetic information to do so, can't produce hemoglobin. Kidney cells express different proteins and functions from heart cells, even though both contain identical genetic instructions. Each normal cell has "brakes": controls that allow certain messages or codes to be opened and translated while others are set aside and ignored. In other words, each cell is able to prevent certain genetic messages from being expressed. Even in the small sampling of examples I just provided, you can begin to see the complexity of the strict rules DNA lays down to create our intricately crafted bodies.

Figure 6.4 The eight trigrams that are part of the *I Ching*. The solid line represents yang, the creative principle. The broken line represents yin, the receptive principle. Each trigram represents different qualities. Two trigrams make up a hexagram.

What tells the cell to turn off a message or express it? This question makes for some of the most exciting research to date. An emerging concept called *epigenetics* indicates that *the environment* can modify gene expression.[8] Epigenetics shows that though the genetic code itself may not be altered, what those genes do can be influenced by the diet, drugs, and even lifestyle. Dr. Dean Ornish explored this idea in men with prostate cancer. Those men who stuck to specific lifestyle habits for three months, including a vegetarian diet, stress reduction, and social support, showed significant changes in prostate cancer genes. The *expression* of the cancer genes was lowered significantly.

Numerous studies in animals reveal that external influences affect what genes will do, and the changes in genetic expression can be passed on to subsequent generations. What a mother eats during pregnancy can affect how the genes are expressed in her offspring. This was first demonstrated in the now famous agouti mice experiments done by researchers Randy Jirtle and Robert Waterland at Duke University.[9] Agouti mice are obese yellow rodents that have these characteristics because of the presence of a particular gene—the agouti gene. When they breed, their offspring not only share these characteristics but are also prone to cancer and diabetes. When the researchers changed the potential mouse moms' diet just prior to conception, their offspring, though carrying the agouti gene, were healthy and lean brown critters. The specifics of the diet ensured that the mother mouse got lots of simple, regulating molecules found in such foods as garlic, onions, and beets. The DNA regulating molecules, called *methyl donors,* include choline, methionine, and folic acid and are available in our food and supplements. For decades, pregnant women have been encouraged to take adequate folate to ensure healthy brain development in their babies.

Epigenetics research is exciting because it provides evidence for an entirely new understanding: we are not necessarily defined by our genes. Devastating genetic illnesses may not have to be passed on to our children if we can learn how to effect the change. Genes are not immutable, or fixed in stone; on the contrary, their expression is actively open to change. Genes are not pulling the strings of our life—instead, we and the strings of our cytoskeleton are pulling them! Essentially, the field of epigenetics shows that the actual structure of the gene or its code is not changed; only its expression is altered.

Pioneering scientist Bruce Lipton tells us that our beliefs and attitudes can also change our genes, as might imagery and other healing practices.[10] Knowing that it's possible to alter our genetic fate may contribute to our ability to do so.

Exploration

Body Prayer: Unwinding Our DNA

You will recognize similar practices from chapters 2 and 5. This time, reflecting on DNA, we emphasize the spiral nature of the movement. Stand facing something you find beautiful, or close your eyes. Feel your feet firmly on the ground, with the soles of your feet in full contact with the earth. Your head is upright and your spine is straight. Place your tongue on the roof of your mouth behind your teeth; this is the "inner smile."

Now begin moving in a spiral by inscribing a circle with your waist, moving clockwise for a few minutes and then counterclockwise. You might even feel your back getting involved in the movement. Breathe easily and see whether you want to spiral more in one direction than the other. You do not need to count your circles; just tune in to the movement. Humming the sound "mmm" while spiraling may be just the key to soothing your cells, your self. As you move and hum, you can also imagine that this easy, stress-reducing, energy-generating practice is correcting any errors in your DNA. When you feel grounded, stop. Feel your feet still in full contact with the earth. Feel yourself rooted in the earth and connected to the sky. Clasp your hands to your belly and give thanks.

You may want to experiment with doing this basic qigong posture in spare moments throughout the day. It builds a bridge that connects the earth, your cells, and the heavens.

Making and Repairing Mistakes

Each day about a hundred billion cells in the human body divide, making new cells. Recall what happens to the DNA as a cell divides—the paired strands of DNA separate, unwind, and then are copied, letter by letter, serving as the template for an identical partner to be created. While this is happening, a mistake can be made by mismatching the triplets, omitting or adding in a wrong letter; such a mistake is a *mutation.* A mutation of a single code letter can change which amino acid is placed in the scripted protein. Even one change can alter the shape and function of the protein being produced—an incorrect code can make the protein stiffer or too flexible or a totally different shape from the original. The protein may no longer work the way it is supposed to. Since even a single, minuscule error such as this can affect the health of the cell, a potent repair mechanism must be available to protect the cell from damage. In fact, in nature's wisdom, multiple repair systems are present in our cells.

At the genesis of this growth of cells, self-correction is insured
by the wavy strands of molecular intelligence held tight
unwinding and letting go only after perfection is created.
—CHRISTOPHER VAUGHAN *How Life Begins*

Much of the time, our cells get it right, yet sometimes they don't. In fact, it's been estimated that at least a thousand errors are committed inside our cells every day. Fortunately, the cell possesses innate wisdom, built into the architecture of the DNA helix, that recognizes when an error has been made. Damage or errors in DNA trigger an astonishing sequence of events as a gene called p53 rides to the rescue.

The p53 system is both a "spell-checker" and an emergency brake on cell growth, and it has other genes under its command. If an error is created, the p53 gene orders other genes to stop being copied until repairs can be made to the DNA. Once the damaged DNA is repaired, p53 turns on the green light and allows the cell reproduction cycle to continue. But what if the damage is beyond repair? In that case, p53 activate genes that direct the cell to self-destruct; this is known as *programmed cell death* or *apoptosis,* which comes from the Latin for "falling leaves."[11] Apoptosis, in contrast to traumatic or necrotic cytotoxic death, is a relatively gentle process in which parts of the cell slough off and are recycled or removed by the scavenger cells—falling leaves are an apt metaphor, with their ability to decompose, be recycled into the earth, and even nourish the tree that once sustained them.

A traumatic or cytotoxic death, by contrast, is one in which the cell is acutely damaged and basically explodes, releasing its contents into the cellular environment. This type of cell death can damage the surrounding tissues as it sets free potentially dangerous substances from the cell. Cells contain numerous substances that, if released, can harm other molecules; however, within the cell they are compartmentalized to protect against their damaging effects. Apoptosis is a slower process that allows the neighborhood to reclaim or eliminate cell parts, one step at a time, without damaging the area.

To sum up the role of p53, it's the damage-control specialist with the capability to correct gene errors, prevent amplification of unruly DNA, suppress tumor cell growth, and when necessary, push cells into programmed self-elimination. Our cells have other numerous backup systems as well to ensure healthy survival.

Death, a Natural Process

Normal cells do not live forever. Death is an ordinary and necessary function of our cells. And the programmed cell death just described is not limited to a last-ditch cure for DNA mistakes. Apoptosis (you could call it assisted suicide) actually ensures normal development.[12] In our fetal phase of life, for example, during the "fin stage" when we have not yet developed hands, programmed cell death eliminates unnecessary cells in the area, and voilà!—fingers are formed. Apoptosis also eliminates any renegade immune cells in our thymus that could mistakenly attack our self cells. Brain cells that don't connect with others are also programmed to die. Cells that are irreparably damaged are expected to commit "hara-kiri" for the good of the cellular community. So within each cell, there must be knowledge it has to die. A characteristic of the abnormal cancer cell is that it has *forgotten to die.*

Taking a moment to reflect further on the normal, continual, gentle dying process of our cells, we find the genetic reminder that the dying process is a natural part of life; it is not a mistake or a failure.

Cells have multiple strategies to ensure elimination of unhealthy cells—p53 genes are not the only ones in charge. A strip at the tip of our chromosomes also regulates when a cell will die (see figure 6.2). Called telomeres, a repetitive DNA sequence (such as TTAGGG) akin to a string of pearls, these are a kind of molecular clock that ensures mortality.[13] Each time a cell divides, it loses some "pearls," which shortens the chromosome. At birth, the length of the telomeres in human white blood cells is about eight thousand nucleotide units long. In the elderly there may only be fifteen hundred units left in a cell. Typically, a normal cell can divide about fifty to seventy times, with the telomeres shortening each time. Eventually, the strip of telomeres becomes so short that the cell fails its internal checkup and is identified as a damaged or senescent cell. Finally, in cellular "old age," it is ready to be eliminated. It stops dividing and dies a quiet death.

Exploration

Journey to the Unknown

Let's stop here a moment to reflect on our own death. In Western culture, death is often seen as a failure rather than a natural result of living. In fact, it's the only thing we can be sure of—life as we know it has a beginning, middle, and an end. We may fear death, yet if we understand it as an inevitable, gentle, and natural process, is it possible we could face that stage of life with more ease and less fear?

As someone who worked with children facing death, I had to look at my own fears, expectations, and beliefs around death. And much of my education in this area came from working shamanically. In 1989 in San Francisco,

I taught a course with my shaman teacher, Tom Pinkson. My responsibility was to teach the science of body-mind; his was to bring in ancient indigenous practices. One such practice was the death journey, something I had experienced with him many times. He prepared the class by warning them a week ahead of time what we were going to do—most were less than enthusiastic about the prospect.

Then a big earthquake struck the Bay Area, throwing off our class schedule, and I had to lead the class on the death journey—a first for me. After we completed the exercise, the discussion was profound. An HIV-positive man admitted he hadn't wanted to do this at all, thinking he was likely the one in the class who was closest to death. But instead of experiencing fear or dread, he found his imaginary death a tremendous relief. Another person said it gave her the opportunity to plan her life celebration party. I offer the exploration to you here, to be done at a time when you can take about a half hour to go deeply into yourself. Though this is typically done accompanied by the beat of a drum, you can do it without in this simplified version.

As with all the inner exercises, put yourself in a safe place where you will be undisturbed. Have writing or drawing materials handy to capture any insights when you're through. Remember: whatever you experience will do no harm.

Sit or lie down, close your eyes, and let yourself tend to your breathing, the rise and fall of your belly and chest, the in-and-out movement of your breath. Feel safe and at peace. Imagine you are in the last few minutes of your life. Where are you, and who is with you? Create this in your mind's eye as you would like it to be. Let yourself be in this place for a few moments.

Your breathing stops. Your heartbeat disappears. Your soul and spirit leave your body. Your consciousness rises up and, looking down, you see yourself and your loved ones with you. You can stay here watching and listening to what happens.

When you feel ready you can move on to your funeral or memorial service. Float into the room and hover above those who are gathered. Where is this, who is there, and what are they saying? Now see your obituary—what does it say? Allow as much time as you need to experience these impressions. When you are ready, bring your awareness back to your breath, your heartbeat, feeling relaxed and full of life. Draw or write what you feel.

If you choose not to go through this whole process, another option is to write your obituary as you would like it to read. Take steps in the present to do what's needed to live as your obituary says you have.

◇◇

Once I discovered that I could lead people in this journey, I began including it in many of my classes. One student, on reading her imaginary obituary, saw that she was described as a writer. She had not been writing at the time and saw that this was something she had to do. Since then she has written two books.

This death journey is an extension in our lives of the lessons contained in our trillion cellular sanctuaries. They can teach us to live in more whole, holy ways, and we can honor them with how we choose to live.

Cellular Glitches

As you increase your knowledge of the cellular world within, you also discover that every normal process has its glitches and that cancer is one such glitch. It's interesting that cancer cells are often called *transformed* cells, yet from the standpoint of consciousness, transformation is generally thought of as something positive. The transformation some people with cancer experience is to make each day count, to reprioritize life to include what's most important. Perhaps transforming cells hold the message to change how we are living.

So how do normal cells transform themselves into cancer cells? Some make a new protein called telomerase that protects the "string of pearls" by adding onto the chromosome tips the repetitive DNA sequence we discussed earlier that "marks time," tracking the age of the cell: those cells then become immortal, relying on telomerase to keep their cellular clocks ticking, no longer shortening their lives (and chromosome tips). The good news about finding an abnormal protein such as telomerase only in cancer cells is that specific therapies can be directed exclusively to cells containing this protein.

Eluding Detection

Cancer starts when something goes awry inside the cell, such as an error introduced into the genetic code, which can occur during normal cell division when DNA is being duplicated. Genetic errors may also be caused by a toxic chemical, radiation from the sun, X-rays, viruses, free radicals, cigarette smoke, or other carcinogenic agents. Altered genetic messages can cause abnormal growth, malignant transformation, and eternal life for the chaotic tumor cells.

A single genetic mutation doesn't turn a normal cell into a cancer cell. Most human cancers develop following multiple genetic errors.[14] The development of cancer over a lifetime results from a series of gene mutations in regulatory genes; regulatory genes include oncogenes and tumor suppressor genes, which stimulate cell growth and inhibit proliferation. Mutations acquired throughout our lifetimes can alter regulatory and communication functions of the cells. This is one reason the cause of cancer is so difficult to determine—it may take decades to generate the number of genetic deviations necessary for cancer to develop. Very few cancers are inherited. Keep in mind that cancer genes, such as BRCA1 in breast cancer, indicate increased vulnerability, but cause and effect are not 100 percent certain.[15] In order for most breast cancers to develop, multiple genetic errors must occur. Likewise, in colon cancer, four separate genetic errors to regulatory genes—in this case, tumor suppressor genes—have been implicated.

In cancer, even though genetic information may be altered, the cell's self-identity markers may remain unchanged. Remember that the physical identity of a cell, its "face," is what stimulates an immune response *only if it is recognized* as an enemy. The immune system can recognize an altered cell or a tumor cell provided there are chemical clues that say "different—not self." And herein lies a problem. The "not self" signals that are so effective at alerting the immune system to act are located on the cell's outer surface; genetic identity, by contrast, resides in codes protected in the inner sanctum of the nucleus. If there are no surface changes or chemical clues, the immune system is not alerted to the danger of impending cellular anarchy. Therefore, the immune system is not the primary defense against most cancer cells; the gene repair system is. Yet in many cancers, it is the DNA repair mechanism that has failed.

Mutations in the structure of p53 can negate its surveillance capability, allowing abnormal cells to develop. In fact, mutant p53 has been detected in more than half of human cancers.[16] Cancer cells that contain mutant p53 have a less favorable prognosis and require more aggressive treatment.[17] Besides making the repair mechanism much less effective, mutant p53 can prevent cancer cells from being killed. Agents that can experimentally push p53 into mutant status include cigarette smoke; smokers who develop lung cancer show changes in p53 not seen in nonsmokers.[18]

Yet there is hope here as well. In people who started smoking as adults and then stop, abnormal p53 disappears. In other words, the damage from smoking is reversible, provided they were over twenty years old when they started to smoke. If they started smoking as adolescents or younger, the gene damage appears permanent.

Many cancer cells won't die; these may have faulty communication or inadequate repair systems or may not be identified as dangerous. Is it possible within the new biology and body-mind wisdom to harness inner healing resources to allay these problems? Can we intervene before too many gene errors are made? One thing we know we can do to protect our genes is alleviate stress.

DNA Repair Rates Revamped

The rate of DNA repair—how quickly errors can be fixed—influences our vulnerability to cancer and other illnesses affected by genetic mutations. Long-term stress slows down DNA repair, as does cancer. In China, a study on improving the rate of DNA repair offers tantalizing and hopeful results.[19] Researchers found that the DNA repair rate of people with cancer in remission compared to healthy people was much slower. The patients in remission were then taught qigong stress-reducing techniques. Following three months of practice, their cell repair rate had nearly doubled.

It is conceivable that the "new" energy medicines of qigong and vibrational sound can affect the erratic energy of DNA.[20] Ancient qigong, tai chi, yoga, and the dance of the whirling dervishes all use spiral movements as part of their energy healing exercises—do they help realign our DNA? Prolonged stress damages the immune system, reproduction, digestion, memory, and even our bones. That is why stress reduction practices, which include the practices just mentioned as well as meditation and imagery, are important to our health at every level.

Imagine That!

The use of guided visualization and imagery is growing in acceptance as a complementary healing modality, particularly in stress reduction, and easing pain, suffering, and other consequences of cancer and its treatment.[21] Significant data indicate that the miserable feelings associated with a cancer diagnosis and the side effects of treatment can be minimized in some people who practice imagery. Many of the first popularized imagery scripts had people visualizing their immune cells coming to the rescue and killing cancer cells. What we have learned since is that the immune cells are not the primary removers of "demon" cells. So what if we instead base our imagery on a new, spiral model of fixing genetic errors? Here I offer two different suggestions for eliminating any abnormal cells in your body. They are only suggestions; feel free to use your imagination.

Exploration

Eliminating Unhealthy Cells

Take some time to relax and pay attention to your breath. Feel all the places your body touches: the chair, floor, or other surface you rest upon. Allow your breathing to be peaceful.

If you know that you have cancer cells in your body, imagine what they look like. Biological accuracy is not necessary—how do you *perceive* them? Now allow yourself to imagine something that will eliminate those cells. For example, you might imagine the cancer cells as dust mites and the eliminator as a vacuum cleaner. Make sure the eliminating force is larger and stronger than the tumor cells.

Once all the abnormal cells are removed, picture new, healthy tissue developing. Take as long as you need for the process and then bring your awareness back to your breath and this present moment.

Repair and Cover Up

Since our cells make errors all the time and abnormal cells do exist in our bodies, this script focuses on changing abnormal genes.

Take time to relax the same way as for the previous exercise. Now imagine or intend that any errors in your genes are corrected. Picture the spiral DNA pairs being made to match perfectly, removing or correcting all the errors in your genetic repertoire.

Alternatively, you can envision preventing these genes from being expressed by covering them up with new proteins that adhere to them, keeping them hidden.

Allow your imagination to guide you in the way abnormal genes are taken out of action. See all of your genes as healthy and whole. Take as long as you need for the process and then bring your awareness back to your breath and the present moment.

Draw or write what you experienced.

The divine choreography of our DNA—its spiraling strands and ability to program both reproduction and self-sacrifice—brings us once more to life and death. Embedded within our cells is the ability to detect and correct errors in coded genetic messages. When correction is impossible, a gentle death is initiated. We know that ultraviolet radiation from the distant sun can penetrate and mutate the gene. And what about cigarette smoke that somehow is breathed into the cell, altering the gene structure so that faulty proteins are made? If invisible agents can initiate damaging changes, can we use the invisible laser of our energy or imagination to cut out or hide the damaged sections? Ancient healing practices including walking the labyrinth and chanting may provide valuable assistance for transforming our inevitable cellular errors.

Body Prayer

Giving Thanks

This is an easy practice you can do anytime. Assume the qigong Standing Home stance and begin spiraling your waist and body as instructed earlier in this chapter. As you spiral, give thanks to all of your cells for carrying you, holding you, and keeping you healthy. You and they are engaged in a glorious collaboration—creating your life. Add any other thank-yous that come to mind.

Life's Purpose

Each cell has a purpose, a reason to thrive while supplying its particular skills to the rest of the cellular community. Each has a place in our unique, invisible universe. We, too, have a purpose: to be alive and to thrive. Eckhart Tolle suggests that we ask, "What does life want from me?" Our cells live for a finite time, and we, too, have only so long before our spirits leave our bodies. What do we need to do before we go?

I believe that we each carry within us our legacy, our reason for being here. Our life's journey is to find out what we are here to create and do the best we can to express it fully and for the benefit of all.

Reflection

Where in my life am I the most creative?

How do I envision my ideal cells?

What is my main purpose right now?

Where do I notice spirals in my own life?

What must I do before I die?

Chapter 7

Memory–Learn

All memories (wounds and joys) are woven . . . in the human biography and attached to qualities of time within the body in the muscles, organs, even the bones, and in physiological rhythms. These rhythms also exist in the vast realms of soul and/or spirit, and are activated by the olfactory and auditory systems.

—THERESE SCHROEDER-SHEKER "Music for the Dying: The New Field of Music-Thanatology," *Advances: The Journal of Mind-Body Health*

The Heart's Tears

Have you ever had to go wholeheartedly into the world when you had a broken heart?

The first time I unwound a pattern of past body memories, my heart was broken. It was filled with rage. I wasn't sure I could cope with the intensity swirling inside and do my work at the same time. I truly wanted to share what I knew, though, and to do so I had to calm my aching heart.

I had been invited to teach an oncology staff the practices I had found useful for healing and reducing stress. It was an exciting opportunity to go into a clinical setting to teach health professionals who worked with patients. The person who had invited me, the oncologist who ran the clinic, had also been my lover—a week before this planned event, we had broken up. Now what? Do I still go, or do we cancel? We talked about it and agreed that we should remain professional despite our personal challenges and go ahead with the program; his staff had been looking forward to it for months.

The clinic was hundreds of miles away. On the long drive I realized how hurt, sad, and angry I was. I checked into a hotel, spent a fitful night, and awoke upset. More accurately, I was an emotional mess, and I knew full well that it's impossible to teach healing methods when you're emotionally distraught yourself. What to do?

It can be easy to reach a peaceful state when your life is relatively calm, but not so when it feels like you're falling apart. It was an evening class, so I had some time, and I used it to try to quiet myself. I took a long walk on the beach. I practiced qigong. I prayed. I meditated. Hours later, I had to admit that nothing had worked. Then I remembered something I had taught to patient groups before but rarely did myself. By then I figured I had nothing to lose.

I spent about an hour doing this practice and felt myself growing increasingly peaceful and centered. It was so effective, in fact, that I actually became excited about teaching again, delivering practical skills that I knew would make a difference to this oncology staff and their patients. When it was time for the class, the calm I had achieved persisted, even, to my amazement, when *he* unexpectedly showed up.

What this experience taught me is that our cells can be brought into a peaceful state even when that seems unlikely emotionally or even impossible. How our cells set up memories within our muscles and our minds is the subject of this chapter.

Having now taught this particular set of practices to thousands of people, many of whom were survivors of 9/11, I know its possibilities and healing potential. I offer it here for you to experience directly before we get into the science of this chapter. Some of us learn best through experience and only then become interested in intellectual discourse. After you have explored this practice, I will break down the instructions and discuss how they affect you at the level of your sacred cells.

EXPLORATION

Gratitude and "Re-minding" Our Cells[1]

Set aside about fifteen to twenty minutes in a safe place where you will be undisturbed. Have a journal or notebook handy to record your experience after this exploration. After you read these instructions, relax and close your eyes to do the practice. You can also read this aloud and record it, or go to my website (sondrabarrett.com/communication-2) to listen to and download the Gratitude exercise.

Find a comfortable place to sit or lie down. Close your eyes. Feel the places where your body touches the chair or floor. Let yourself be held by the surface you are sitting or lying on. Become aware of your breath as it moves in and out of your nostrils. You might notice a difference in temperature as it flows in and out. Follow the rise and fall of your chest or belly, noticing how deeply you take in each breath. Become aware of its rhythm. Stay with the rhythm of your breathing for a few moments before continuing.

In this exploration, you will imagine, sense, or even pretend that a particular place in your body is breathing. Stay in each place with your awareness until you are ready to move on.

To begin, bring your awareness to a point on your forehead and imagine, sense, or feel as though that spot is breathing.

Next, bring your attention to a place between your nose and lips and imagine or sense it breathing. Now a spot on your chin is breathing. Now a place on your neck or throat area. Next, bring your attention to each of the following parts of your body in turn, staying at each place until you sense that spot breathing:

A spot on each of your shoulders

The inside of your elbows

The palms of your hands

The outside of your ankles

The soles of your feet

Your inner ankles

Your calves

Behind your knees

Your inner thighs

Your belly—feel your belly breathe

Next feel your rib cage move as you breathe. Feel it cradling your heart.

Bring your awareness to your heart and sense or feel your heart breathing. Feel your heart *beating*. You may even be able to hear it. Once you have a tangible sense of your heart, remember *with your heart* a person, time, or place for which you felt grateful, or a time when you were appreciated by someone else. Receive whatever image or experience comes first, without judgment. Allow this image of gratitude or appreciation to become as real as possible. Notice where you are and who is with you. Notice any smells, sounds, and feelings; make this a full sensory experience so that it feels real.

Your heart is remembering gratitude.

When you are ready, imagine with each beat of your heart that your *heart cells are sending this memory of gratitude to all your cells.* With each beat of your heart, that message is broadcast up into your shoulders, along your neck, and into your brain. With each beat of your heart, the experience of gratitude is sent down your arms and legs, into your back and belly, and down to your feet. Allow this experience to *resonate throughout your body, mind, and cells.* Stay with it as long as you like.

You may also imagine sending out gratitude and appreciation to someone else if that feels right to you. You could even send the experience of gratitude to all who occupy the earth with you.

When you are ready to leave this place, *anchor the experience* by gently touching your thumb and index finger together on each hand, making circles. You may add another sensory anchor such as a specific scent or sound. Any of the five physical senses will help weave the memory into your cells.

Now bring your attention back to your breath and the room. Become aware of the chair and the floor; notice how you feel. Let your fingers relax and open again, releasing any other sensory trigger you've used as a memory anchor. Open your eyes. Notice how alert and refreshed you feel, and know that you can recreate this experience anytime you need to.

Take a few minutes to write, draw, or physically move to express and reinforce this experience. This will further anchor it into your cellular memory.

◇◇

This practice is a wonderful way to shift both attitudes and physiology. Each time you repeat this experience with your sensory anchors, your cells are learning. Soon it will become easier to reverse your frown or down mood. After a few practice sessions, your sensory anchors alone may be able to bring you to this safe and tranquil place.

Your Body, Breathing

Now let's explore the steps and the science underlying the practice. The first part, breathing your body, is adapted from yoga, hypnosis, and progressive relaxation. One goal of this part of the exploration is to move you into a relaxed state so your imagination can become more active. The instruction—to imagine, sense, feel, or pretend that a spot on your body is breathing—allows your judging mind to get out of the way. You may find that you're saying to yourself, "But these places don't breathe," and that's OK; you can pretend. You can imagine it happening in your mind's eye. The intent is to provide a route to deep relaxation and a tangible, embodied experience. You may discover that doing only this portion relaxes you. I have seen this happen for many people, including myself.

The Heart of the Matter

By the time you get to your heart, this should be a felt-body (kinesthetic) experience. You really can feel your heart. Again, one of the goals is to help you relax and be in your body. Now, the instruction *remember, with your heart, gratitude or feeling appreciated* is emotionally loaded. You may never have felt appreciated by anyone. But you have surely appreciated others, or a beautiful sunset, a wonderful meal. You may remember and feel gratitude for a loved one who has died. And you may experience absolute joy remembering a treasured moment. Whatever image comes to you is what you are feeling the most grateful for in the now, and it may surprise you. Notice that the instructions are to remember *with your heart* rather than to simply remember. You will learn more about the heart's mind further on.

Imagination is more important than knowledge.
It is the preview of life's coming attractions.
—ALBERT EINSTEIN

Imagined Realities

It may seem that imagination is not real—it means making things up, right? But let's look at some evidence to the contrary. Positron-emission tomography (PET) scans have been used to examine which part of the brain is active when someone is engaged in a particular activity.[2] Before the scan, individuals are given a radioactive form of glucose, the brain's fuel. When an area of the brain is being used, and thus firing, it consumes this radio-labeled glucose, and the brain scan lights up that region. In one study, people were told to look at a picture, and specific visual areas of the brain lit up. Then they were told to close their eyes and *imagine or remember* the picture—and the same parts of the brain were illuminated. The brain responds in a similar fashion whether you are imagining or actually engaged in the activity. Numerous studies have repeated these results.

Now for a bit of brain geography. The areas above and beside your ears, the temporal lobes, process sound and memory. The brain span that crosses over your head from ear to ear holds movement memories and sensory impressions. At the back of your brain, the occipital region processes visual inputs. Each sensory strand of information is woven together so that the many sensory threads of the entire experience create a memory or imprint, perhaps holographically. The more times we visit this constellation of sensory memories, the stronger a tool it becomes for healing. And this is true whether our experience is real or imagined.

If you've ever done guided imagery, you may recall that you were encouraged to engage as many of your senses as possible, and the practice you have just explored includes an instruction to remember your gratitude experience with as many of your senses as you can. This is to take advantage of the collection of sensory memories just described. The more times we travel a road, the easier it is to find our way.

Our Senses as Doorways to Memory and the Sacred

Our most primitive sense, smell, also is the primary initiator of memory. Be it the smell of a roasting chicken, pizza, perfume, or incense, by taking in the scent, we remember the past or a sacred moment. Remember that newborn babies can recognize their moms' smell within their first twenty-four hours of life. Sound, too, can set our cells vibrating with a remembered experience or a sacred hum or hymn. And we can rely on our physical senses to teach the bodymind new behaviors. The senses connect inside and out, allowing us to take in the world and open to its blessings.

Anchoring and the Senses

To complete the imagery, the instruction is to touch your fingers together and/or use a sound or scent to anchor the memory, another invitation to engage multiple senses in training the cells to remember. Conditioning—or as I call it, *reconditioning*—our cells is a powerful healing strategy that is always available to us.

Writing, drawing, or moving after the experience helps even more cells connect and remember. With any relaxation or healing practice, adding a sensory anchor encourages the cells to collaborate and build connecting points with each other.

In one of the early Western medical studies teaching meditation, people were given a scent to anchor the relaxed state when they reached it (in this study, lavender). Over time, the scent of lavender alone could initiate a relaxed state. When I did a lot of teaching around the country, driving to a new city after a full day of teaching was often stressful. I'd always have lavender with me. Sniffing it would help me relax and remember to breathe and help release the tension from my shoulders.

One of the instructions for remembering the experience of gratitude was to imagine with each beat of your heart that cells throughout your body are resonating in that state. Remember how the strings of your cells vibrate with sound, energy, or thought? One cell may set the neighboring cells into humming at the same resonant state.

Definition

Resonance has many definitions. Resonance is the quality of a sound that stays loud, clear, and deep: an intense prolonged sound produced by sympathetic vibration.

Another definition is a sound or vibration produced in one object caused by the sound or vibration produced in another. Resonance as sound includes frequency and tone. Frequency refers to a rate of vibration, a pulsation of waves; a sustained frequency is a tone. Each sensory experience has its own tones; touch, aromas, and sounds communicate information at different rates of vibration.

José Argüelles, author of *The Mayan Factor,* describes resonance as the quality of sounding again. To resonate is to reverberate, which implies communication, an exchange of information. Resonance is information.

Resonance is also a quality that makes something personally meaningful or important.

Cellular Resonance: Entraining Knowledge

According to Argüelles, only when we engage all the sensory information from the past can we actually resonate with the experience in this present moment. All our cellular strings hold our memories within us.

In *The Silent Pulse* Aikido master/educator George Leonard recounts the tale of a sixteenth-century clockmaker who built beautiful wooden grandfather clocks. One day when hanging a new clock in his shop, the clockmaker noticed that each pendulum of every clock swung with its own independent rhythm. Then, suddenly, everything changed—all the pendulums began swinging together.

Almost magically, on their own, the clocks will begin to beat in synchrony, resonating with each other. When I tended heart cells in a petri dish, each cell beat a separate rhythm until they grew close to each other. Then, like grandfather clocks in a shop, all the cells beat as one. This phenomenon is also known as *entrainment.* Now consider that our grateful, beating heart cells can coax the other cells into line, all in the same rhythm. We've ignited our core resonance, and may even be aligning with the universal core of energy.

Cells entrain too as the field of energy changes. Biologist and futurist Rupert Sheldrake would say that our cells, when entrained, share a field of morphic resonance. The field of energy surrounding these cells tunes them to the same vibration.

You can enjoy another experience of resonance or entrainment by listening to music with a slow beat and then with a fast beat. Notice what happens to your heartbeat. Or what about listening to music that makes you happy? For me, while writing this book, Jason Mraz's "I'm Yours" embraced me with joy, always. When we resonate with a piece of music, an idea, a place, or a friend, we echo and reverberate the same "vibes."

Why Gratitude?

When I first began doing imagery with people, I gave the instruction to remember a peaceful or joyful time. I was shocked to discover that some people had no memory of being peaceful or happy. Gratitude seems to be much more universal and achievable. We may feel gratitude for a person, a place, a movie, a pet, or something someone did for us. We may have been acknowledged and appreciated by another person. There are lots of reasons to be grateful. We may even feel gratitude for being able to pretend to be grateful! Gratitude journals have been made popular by Oprah Winfrey and Brother David Stendl-Rast. Books on gratitude and research on its benefits abound.

Gratitude is an attitude.

—CAROLYN MYSS

A growing body of literature in the relatively new field of positive psychology shows that keeping a gratitude journal or taking the time to remember three things that occurred in the day for which one feels grateful benefits physical, emotional, and social health. Heart rate and blood pressure are lowered. Immune health is enhanced. People become kinder, more generous, and more empathetic. When gratitude or appreciation is expressed in the workplace, people are more cooperative and productive. It certainly can't hurt any of us to take a few minutes every day for gratitude. Our very cells will be grateful if we do.

I recently taught a weekend Cells and the Sacred workshop. Though I am in the habit of occasionally thanking my cells for taking care of me, on that Sunday morning in the garden I was truly grateful for everything my cells bring me, including the wisdom I share here. I felt a response inside, a giggling and a message of gratitude that said, *Finally, you got it—you really* mean *it.* Were my cells talking to me? Was I simply resonating in a state of remembered gratitude and my energy inside ignited with joy? All I know is that my relationship and communication with my cells has deepened since then. We play better together, now more than ever.

Gratitude and the Heart

According to investigators at the Institute of HeartMath, what we feel and remember influences how our heart cells beat and project electromagnetic energy.[3] When we feel positive emotions, the interval between heartbeats differs from the rate when we feel fear or anger; this interval between heartbeats is called heart-rate variability (HRV).

HeartMath researchers found that increased HRV is associated with more positive emotional, mental, and social interactions. When we are in states of anxiety, anger, or fear or are holding "negative" thoughts, our heart generates decreased HRV. When we stop a "negative" thought or feeling by remembering a moment of appreciation or feeling loved, HRV increases. We radiate a field of electromagnetic waves of energy that influences those around us; the magnetic field around the heart varies with changes in HRV. When hearts and minds are peaceful, people resonate positively with one another. In fact, businesses that teach their employees how to shift mental and emotional states report improved workplace cooperation, relationships, and productivity. The work of the Institute of HeartMath illustrates that our heart cells are powerful generators of resonating energy. Some attribute this to the *brain or mind of the heart.*

The Heart's Mind

Psychologist and author of *The Heart's Code,* Paul Pearsall was one scientist who believed that heart cells do in fact have a mind—and with good reason. While he was in the hospital after receiving a bone marrow transplant, Pearsall heard intriguing stories from other transplant patients, particularly those who'd received new hearts.

In one remarkable story, a middle-aged woman behaved very differently after receiving the heart of a twenty-year-old biker. A lover of classical music, she now listened to rock and roll and craved green peppers, junk food, and beer—all of which had been passions of the biker.

Another tale is even more compelling. While driving down a dark country road, a husband and wife were arguing. A truck swerved and hit their car, and the husband was killed instantly. Hours later, his good heart was transplanted into a young Hispanic man. Months passed. The wife, who was being counseled by Pearsall, asked to meet the recipient of her husband's heart, and it was arranged. She asked the young man if she could place her hand on his chest. He understood her need to connect, and agreed. With her hand on his chest, she said, "Everything is copacetic." The recipient's mother gave her an odd look and then asked what "copacetic" meant. The woman told her it was the word she and her husband would say after an argument when everything was OK again. The mother smiled. "That's the first thing my son said when he came out of surgery: 'Everything's copacetic.'"

Mysterious? The young man spoke little English—how did such an unusual word come to his mind? Can heart cells truly carry such memories?

Pearsall reminds us what other cultures say about mind: it does not reside in the brain; it dwells in the heart. The Chinese characters for heart and mind are the same. Think with your heart, our Native American relatives tell us. If your heart were given to another, what memories would it carry to them? What memories would you want it to offer?

To most scientists, the idea that the heart—or a cell—has a mind or memory seems absurd. Mind resides solely among the neurons in the brain, they believe. And indeed, for more than a hundred years scientists have sought the seat of memory within the brain. Yet according to some scientists, rather than being localized in one place as originally theorized, memory and the mind are everywhere within us.

Where Is Mind?

Does the capacity to learn reside in the mind, brain, or the whole body? Where is the mind? The physical brain of course, unlike the mind, has a distinct location in the body. The brain is physical; it can be touched, dissected, probed, and measured. Encased within the stronghold of the bony skull, our brain, a three-pound wrinkled hunk of wet tissue, is certainly capable of learning. It can transform multiple molecular messages into electrochemical signals and meaningful patterns of information. A vast network of more than 100 billion cells helps direct the neurochemical traffic of incoming data and outgoing instructions. The network carries instinctual orders for our survival by maintaining the basic operating system of our body: breathing, heartbeat, temperature, the instinct to eat and drink, and so forth. Though it operates by instinct, the brain can also be trained and programmed with patterns that keep us alive with pleasure.

The mind, on the other hand, has no tangible dimension as we know it. We can't touch it physically; it's invisible. We can only measure its effects. Yet the mind and brain must interact. The brain carries out the mind's wishes, and together they learn. A single neuron does not learn by itself. Learning takes place in communities of neurons, and memory is the network of cellular and energetic connections. Just as described earlier in this chapter, our sensory channels create a network of each specific experience. According to Stanford University neuroscientist Karl Pribram, memory is stored in wave forms throughout the brain and body. Once more we meet the idea that vibrational waves of energy carry information.

One Piece Remembers the Whole

A holographic image captured on film is unlike a regular image. Shining a laser through any piece of holographic film can project the whole image rather than just a portion. Pribram says that memory works the same way, that each memory is holographic, held in a code of waves throughout our entire body.[4] Activating one piece of the holographic impression triggers the rest; one part remembers, bringing all the other connected parts into play.

When I remember my son or daughter's face, their actual face, or even a picture of it, is not residing somewhere in my brain. Yet how can I see them so clearly when I think about them? Somewhere in my body is a code for each face. Our mind-brain holds information somewhat as a computer holds its information—in a patterned code. How else could we carry a lifetime of memories, millions of cellular skills, and a thousand ideas if these were not somehow compacted into an electrochemical coding system? Though Pribram's notion of memory and mind is not universally accepted, he provides yet another argument that our cells know more than we think they do and that their intelligence is carried in waves of vibration.

Deeper into the Construct

Sir John Eccles wrote that the electrical exchanges between brain cells (synaptic potentials) don't occur singly. Every nerve has branches. When an electrical message goes down the branches, a ripple or a *wave front* is formed. When other wave fronts come to the same location from different directions, they intersect and set up an interference pattern. It's like the meeting of ripples that form around two pebbles thrown into a pond.

So according to the holographic theory, our memory is constructed from vibrational patterns and becomes activated when the right set of wave forms is transmitted. When a song or the scent of newly mowed grass floods us with memories, it's because the wave patterns trigger a set of stored holograms. What we call *situational cues* for memory are none other than a set of wave forms that can activate the appropriate hologram.

Perhaps herein lies the answer to the mysterious effects of the transplanted heart; it carries holograms of wave forms that inform its new owner of preferred tastes and sensibilities.

Though there is no objective proof of this hologram theory, it carries a compelling idea; cellular memory and intelligence still remain enigmatic. Whatever it is and wherever it resides, mind is an enigma. Yet as

I have become aware of this hologram theory, I am discovering that as I attempt to let go of a memory, I can imagine waves leaving me.

> *What goes into the making of the neural holograms that the brain uses to experience reality are the images upon which we meditate, our hopes and fears, the attitudes of our doctors, our unconscious prejudices, our individual and cultural beliefs, and our faith in things both spiritual and technological. These are important clues that point towards why we must become aware of and acquire mastery to unleash these talents.*
>
> —MICHAEL TALBOT *The Holographic Universe*

On One Condition: Sensory Learning

A person with cancer standing across the street from the hospital where he received chemotherapy the month before feels the same discomfort of the previous treatments by simply looking at the front door. The smells inside the hospital further remind him of his stomach's queasy reactions. Seeing the building and smelling the odors associated with his upsetting experience, he is being unconsciously conditioned, his memories and nausea rekindled.

The cellular body-mind recalls sensory conditions when it learns. Psychologist Ernest Rossi calls this *state-dependent learning.*[5] Pribram would name it *holographic.* From the multidimensional pictures in your mind—the smells, sounds, feelings, sensations—you create a holographic sensory state associated with a previous encounter. Any one of the sensory triggers can bring back the whole experience. We learn conditions "conditionally."

Does the Name Pavlov Ring a Bell?

While sensory input can trigger an old memory or prejudice, it can also help us learn new behaviors, like feeling gratitude. Remember Ivan Pavlov? Experiments more than a century ago by the Nobel Prize-winning Russian scientist led to the discovery of behavioral conditioning.[6] Dogs automatically salivate when they see a piece of meat. The sound of a bell doesn't have the same effect; it will not trigger that autonomic physiologic response. Yet Pavlov trained dogs to salivate at the ringing of a bell. How? Each time he showed meat to the dog, he also rang a bell, and over time, the sound of the bell even when meat was not present caused the dog to salivate. In this way, Pavlov discovered that the brain could be changed to acquire new knowledge and behaviors. For his dogs, a bell foretold the arrival of food. The dog's cells learned to respond to unusual triggers, and their physiologic networks connected and learned.

What does Pavlov's discovery have to do with us? It helps us understand how many of our unconscious behaviors and attitudes have been programmed and perhaps can be unlearned and new physiologic patterns learned. Dr. Robert Ader, world-renowned psychologist at the University of Rochester, showed that our immune system could also be conditioned through sensory stimuli.[7] At the time, he wasn't interested in either the senses or immune function—he wanted to understand what it is that builds the memories that cause persistent reactivity to an unpleasant experience. Why does a person who has undergone chemotherapy one time react so drastically the next time he or she sees the doctor's office, hears the nurse's voice, or smells the waiting room?

Ader's objective was to discover how an unpleasant experience programs or conditions the same experience to recur. To investigate this, he first worked with animals. He injected mice with a chemical that made them sick to their stomachs while at the same time giving them a unique taste: saccharin-sweetened water. So here we have a new taste for the animals paired with a one-time unpleasant experience—nausea. Over time, he followed whether the animals would avoid the water or keep on drinking it. Surprisingly, the animals that drank the most saccharin-sweetened water started dying of infections. Unbeknownst to Ader, the drug he used to cause nausea, the chemotherapy agent Cytoxan, also suppressed immune function. The animals had received the chemotherapy drug only once, yet each time they drank more saccharin-sweetened water, their bodies "remembered" the immune-suppressing effects of Cytoxan. So here is an example of the sense of taste triggering a long-term effect. It turns out that taste often stimulates the most rapid conditioning, as other sensory input like touch or smell requires numerous introductions to condition the body. What Ader unwittingly discovered was that immune responses could be conditioned or "trained" through the senses!

Until Ader's work and his subsequent collaboration with immunologist Nicholas Cohen, the common consensus was that the immune system learned only through its attraction and response to foreign, or "not self," antigens.[8] They broke through that long-held scientific dogma by showing that the immune response can be influenced by conventional sensory conditioning. This also opened the door on the role of belief, expectations, and the mind on immune functions. Later, other investigators demonstrated that immune memory could be enhanced or diminished by pairing an immune stimulus with any of the five physical senses. So cells can be educated by our senses.

Exploration

Sensory Conditioning

Sensory conditioning is a useful strategy to put into your "medicine bag" of health remedies.[9] The next time you meditate or do the gratitude practice in this chapter, or are simply feeling relaxed, enjoy a whiff or two of one of your favorite scents, or a new one. When you feel that way again, sniff away. It might take four or five "cellular sessions" to take effect, but soon the scent alone will give you that same relaxed, peaceful feeling. This is a great backup for those times you want to change your mood and mind.

When setting up a pairing experiment such as this, make sure you pair the sensory input (the scent) and the physiologic trigger (being relaxed) within thirty seconds.

Imagery: "Re-Minding" the Body

When I worked with children I believed imagery worked, and taught it for years, but I never used it personally. (Unless we include the imagery associated with worrying, at which I was a champion.) About twenty years ago, when preparing to teach my first course in psychoneuroimmunology, a rash erupted on my left arm. It persisted for months. I changed my soap, body lotion, detergent—nothing made a difference. Cortisone cream didn't diminish it, nor did any other remedy I tried. Then one day I thought, "Well, if you believe that imagery works, why don't you test it out on your rash?" I closed my eyes, breathed deeply for a few minutes, got relaxed, and suddenly an odd image—one a child might imagine—leapt to mind: an elf. This little creature called himself Mortimer, and he carried a gold spike, which he used to clean the rash from below the surface of my skin. When he was through I thanked him, asked if I could call on him some other time, and opened my eyes. The rash was still there—but the next day it was gone.

A miracle? A coincidence? Was it my wish, my belief, my intention that eliminated the rash? I may never know, but my arm stayed clear for weeks. That is, until a former physician colleague offered to share the materials on imagery and body-mind health he had gathered through the years. He had labeled the file "Quackery." The next day my rash returned!

This story serves to demonstrate that not only do our own beliefs influence our healing; we are influenced by others' beliefs as well. My mind both believed and doubted what had happened to the rash using imagery, and when an "expert" suddenly reinforced my doubt, the magical healing process was reversed. Can reinforcement or contradiction of one's beliefs by others explain why some people who are told they have six months to live die in six months, while others live far longer?

We often think that what we imagine is not "real." How, then, do we explain the placebo effect? A medication that is really an innocuous sugar pill can relieve pain, change the course of an illness, or even stimulate the nausea of a chemotherapy agent. Our minds and our cells collaborate in such responses.

> *This holographic matrix provides each cell with its cellular mind. The discovery that every cell contains a reflection of the whole has given us a tremendous clue to the way in which this mysterious, multi-level body is constructed.*
>
> —JOHN DAVIDSON *The Web of Life*

The Mind of Our Cells

The mind is said to live in every cell, and if you change the mind of one cell, the rest may move to resonate with it. Picture the seed of an image beginning somewhere in your brain. This image sends impulses to numerous sensory neurons linked to various regions of your body landscape. Trigger one, and they may all respond, and the potential is truly astronomical. You have at least 10 billion neurons, each with the capacity to make more than five thousand connections. The more links and connections you make through imagery, thoughts, and feelings, the more doorways to change you can open. The more senses and emotions involved with the image you hold in your mind's eye and in your cells, the more pronounced are the electrical, energetic, and chemical responses and information flow throughout the body—and the more powerful the remembering.

When you imagine engaging in an athletic activity, you are "re-minding" the body of its neuromuscular patterns.[10] Olympic athletes from the former Soviet Union perfected their performances using imagery. A surprising story of an elderly yoga practitioner bears repeating here.

George, who had done yoga for nearly ten years, slipped and fell. Now with his leg in a cast, he was unable to continue his practice. Every day, however, at the same time he usually did his poses, he did them in his imagination instead. Usually when a limb is in a cast for a period of time, it loses muscle and strength, but when George's cast was removed, his leg was quite healthy. Was it imagining his cells that helped maintain strength and tone?

My qigong master, DaJin Sun, tells of his own remarkable recovery. When living in China as a teen, he had carried huge sacks of rice until he suffered an accident that broke his back. This is a terribly immobilizing injury. But as he lay in a hospital bed, he began recalling the qigong exercises his mother had taught him. He had never actually done them before, but now he began—in his imagination. When I met DaJin about twenty years later, he was a robust, active man with no traces of his earlier injury.

Remembering to Remember

A teenage boy lying in bed overheard the doctor whispering these words to his parents in the next room: "He won't be alive in the morning." He didn't want to die and pleaded, "Please let me at least see one more sunset." Hearing him, his mother tiptoed into his room.

"Mom, would you please move my dresser?" he asked.

It was a strange request from her very sick son, but she complied. He told her how to place his dresser so that the mirror on top of it would reflect the setting sun. For the next hour, the setting sun was all seventeen-year-old Milton saw.

The next morning Milton was unconscious. And the next. And the next. When at last he woke on the fourth day, he was almost completely paralyzed. All he could move were his eyes and his mouth to speak—barely. It was polio, and for all he knew, this was how he would spend the rest of his days.

His body was crippled, but fortunately, his mind was not.

Curious and astute, he played mental games. Whose footsteps did he hear coming from the barn? What kind of mood was the person in? He listened to all the sounds around him and invented stories about what he heard. One day his parents left him alone in the middle of his room, tied into a rocking chair so he wouldn't fall. Milton looked longingly at the window, wishing he were closer to it so he could gaze out at the farm and the sunshine. Then his chair began to rock slightly. *What had just happened?* Was it the wind, or had his wishing to be closer to the window stimulated body movements he had thought himself incapable of?

For most of us, this experience might have gone unnoticed, but Milton couldn't let it go—in fact, it propelled him into a period of profound self-discovery. Could he imagine moving and make it happen? Could he wish his body into motion? Could it remember what it had once been able to do? Searching his memory for sensations and images of the movements he had most enjoyed, he imagined climbing a tree like a monkey. How did his hands and fingers grasp the tree limbs? What did his legs do to scamper up the trunk and reach a higher branch? Often when imaging these movements, he stared at his hand, and one day his fingers began to twitch. He continued his mental exercises, and he also studied his baby sister, who was just learning how to walk. He studied how she did it. He watched, and he remembered.

Every experience, real or imagined, is remembered in sensory fragments.

Gradually this determined young man became stronger. Movement returned to his body. In less than a year, he was walking with crutches. Though he had been fully crippled at seventeen, the following year he made a courageous solo trip by canoe into the wilderness.

What was it that helped this inspiring young man to recover? Was it only a matter of time? Did his intense desire and his ability to thoroughly recall sensory memories actually enable him to move again more quickly?

So profoundly was he changed by his experience that Milton went on to become a physician and psychiatrist. One of the first to use imagery in medicine, Milton Erickson is now considered the father of medical hypnosis. His awareness, persistence, and imagination changed more than his own life; he has helped millions of others.[11]

There are numerous stories about magical healings in the health literature I could have chosen to relate in detail in this chapter. I offer this one because I find it an inspiring illustration of the profound connection between our cells and our minds—a testament to the power of imagination and intention to change the physical form and functioning of our cells.

Numerous studies with athletes have shown that those who combine imagery with their physical practice improve more than those who avoid the imagery training sessions. One of my students, a fitness coach, tested this out with several of his clients. He had one set imagine lifting weights before actually doing the exercise; they gained strength more rapidly than the clients he trained in the usual fashion without imagery.

Imagery and Healing

The psychologist, the religious believer, the medical scientist, and the mystic or the shaman each offer a different explanation of how imagery works to benefit health. The psychologist may say we are changing our conscious thoughts to align with deeper, purpose-driven unconscious beliefs. The religious person may say God is hearing a prayer and answering it. The scientist may say that imagery is changing the neurotransmitters, molecules, and cellular networks of the body. The mystic and the shaman may say that we're changing the energy surrounding the situation and that this allows change to take place. In spite of the variety of theoretical constructs about the mechanisms of the imagination, how and whether it will work as we wish it to are part of the great Mystery. We can say it works by giving people hope, a sense of control with their lives or their illness. It's a coping skill that can be brought to bear on day-to-day problem solving. One thing we do know is that a state of relaxation is the first prerequisite; the stress-reducing biochemical and energetic environment induces healing through autonomic nervous system balance.[12]

The language of the body as well as the spirit—image and symbol—feeds our healing power. Now, through the doorways of science, we are beginning to understand how the imagination can have such a profound effect on the body.

Imagery helps create connection to the sacred, the self, and the cells.

Forgetting

So now we've learned a bit more about building cellular memory, but how do we diminish the power of a negative memory on our body and mind? Can we loosen its hold on us?

New Age thought has it that there are strings or cords of energy that bind us to another person in our life, both figuratively and literally, and that there are ways to cut those strings if we feel we need to. We can perform a ritual with the intent of disconnecting from a past relationship, for example. I have certainly experienced the power of this practice; the results were short-lived, but nonetheless, something in the ritual enabled me to feel free of the relationship for a while. The memories and longings that had seemed to consume me no longer occupied my every waking moment.

Shamanic practitioners may tell us there is an assemblage point in our body or energy field that holds tight to each memory. Some say this point is between or behind our shoulder blades. We may even feel places in our bodies that resonate with grief or anger, a remembered insult.

Yet as we have learned in this chapter, our memories and obsessions are not linear strings; they have more dimension than that. They are holographic waves that ripple through our bodies. So if we want to forget—or perhaps more accurately—to change the effect of a memory on mind and spirit, the question becomes, "How can we change the pattern of that memory's waves?" Electrical messages between brain cells travel in ripples or wave fronts, and when wave fronts meet, they intersect like multiple ripples in a pond. If we want to diminish the waves of an unhappy relationship from the past, why not, then, introduce new waves that are slightly out of sync with them? In other words, generate a new set of waves to "wash out" or diminish the old pattern. Is this possible?

I meditated on this question with the aid of a favorite Tibetan "singing" bowl. Tibetan singing bowls can be made to "sing." Just like making a wine glass hum by rubbing the rim with your finger, when you rub the rim of a Tibetan singing bowl with a mallet, it vibrates and emits its characteristic sounds. Traditionally, these bowls are made from a combination of at least seven metals so that they produce multiple harmonic overtones. For centuries they have been used in Buddhist rituals, for meditation and for healing. Rubbing the rim and sounding the bowl repeatedly always takes me to a deep meditative state. Remember that sound represents waves of electromagnetic energy. If you ever have a chance, explore how resonant sound and the power of these bowls relax you and carry you into a meditative state. Waves of energy fill you.

In this case, using it to help me find an answer to a question, I rubbed the edge of the bowl with the mallet until the sound waves filled me, and then I rang it again and again. Each time I struck it, I held the intention that these new sound waves would supplant the waves of an unpleasant memory—and I felt a shift. I will continue to experiment with the theory that filling the body with new waves of energy can mitigate the power of holographic memories that no longer serve us; already, in my inner work, I can see and sense the undesired waves leaving me.

I know this sounds like an unlikely solution, and it may be. But if what we think we know about memories being held in waves and vibrating strings is true, will future healers help tune us in to a new state; can we become more in tune with what our cells want for us and leave the menacing magnets of the mind behind? Each of us must find our way toward eliminating what no longer nourishes us and strengthening what does.

Memory, Ritual, and the Senses: Doorways to the Divine

Sound, smell, and our other senses are doorways to invoking a sense of the sacred. They are part of most ritual and spiritual practices. The sound of bells or a chant and the scent of burning sage are among such ways to invoke the sacred. When we attend a ceremony in a church or outdoors in a tent, we may hear chants and songs we've heard before—or some we have never heard but that seem familiar somehow. Waves of incense embrace us. Burning candles and their warm light surround us with a sense of holiness. And through our senses we remember. Our consciousness opens to an "extra-ordinary" state. The whole experience is carried in waves of information encoded in holograms of memory and vibrating strings. We can reignite the light of a sacred ritual using any of the senses through which we experienced it.

In fact, I suspect that one of the reasons religious traditions ritually repeat the same prayers is to reconnect us with our inherent holiness, to help us remember our sacredness. When our ancestors developed these traditions, they surely didn't think about how they were impacted physiologically—such analysis was irrelevant. Still, they embodied the experience. They saw and knew the results, and so can you.

Exploration

Our Senses as Doorways to Heightened Awareness

Our senses are conduits to receiving information, and when we attend to them individually they also serve as doorways to the sacred. I first learned this meditation from Jürgen Kremer at the California Institute of Integral Studies and have since seen it described in numerous spiritual texts.

As always, set aside about fifteen minutes or more to do this in a safe, nurturing space. Read through the directions and then act on the instructions.

Close your eyes and listen to the sounds all around you. Receive the sounds without naming them or looking for them. Be receptive to all that you hear.

You can also do this listening when out in nature, simply walking and listening. You'll be surprised at all that comes to you. The listening meditation is a wonderful way to connect inside and out, matter and ethereal. Try this for thirty seconds at first and then for longer periods of time.

Should you choose to go beyond listening, feel your body breathing. Feel the air touching your skin. Stay with this sense of touch, where your breath touches inside and where you touch the chair, the ground, and so on. Next allow yourself to be open to seeing without labeling or judging. You can simply focus on one of the sensory impressions or stay with listening. As you become adept at this simple, quieting meditation, you may discover you are becoming elevated in mind and spirit. By increasing our level of awareness, we change our consciousness and our internal state of affairs.

Reflection

What do I need to remember?

What do I want to forget?

What reactions or responses are too habituated, too hard-wired within me?

What does my heart's mind desire?

What new wisdom do I want to explore?

What am I most grateful for?

What do I want to remember more often?

Chapter 8

Wisdom Keepers–Reflect

For the ice age cave paintings through the middle ages, art was an expression of our faith that the universe is spiritually coherent. Indigenous cultures lived closer to the chaotic resonances of nature in which the spirit of life was revealed.

—JOHN BRIGGS AND F. DAVID PEAT *Seven Life Lessons of Chaos*

Thus far on our journey we've met our cells in many dimensions—as miracles of molecular construction and as sanctuaries, listeners, messengers, and choosers. We've learned how they function, identify themselves, communicate, learn, and remember. And we've learned a number of practices, inspired by our cells, to enhance our well-being. Each step along the way we've glimpsed our cells' sacred nature.

In this chapter we go a step further and pose a question: "Could cells have served our ancestors in their quest for spiritual knowledge?" Might our ancestors have traveled within and found mythic dimensions in the structure and functioning of their cells?

Is Our Cellular Biology a Doorway to Spiritual Wisdom?

In broaching this question, I am not asking whether spiritual wisdom or longing might be hardwired in our brain cells. Rather, in my speculations I'm seeking to understand whether visioning our cells and molecules could have been a source of sacred wisdom. I am wondering if embedded in the architecture of the cell we might find frameworks for the perennial spiritual teachings passed down cross-culturally through the ages. This does not mean that our ancient ancestors labeled what they saw as a "cell" but that they inherently knew they were visiting their own inner world.

The idea that inner vision and art can precede scientific discoveries is not unprecedented. Leonard Shlain writes in *Art and Physics* that artists "saw" and expressed concepts that we now consider the purview of the physicist long before scientists discovered them. Presaging the science, they were seers, visionaries.[1]

In his groundbreaking explorations of the shamans of Peru, anthropologist Jeremy Narby provides evidence that their visions of the "cosmic serpent" corresponded strongly to molecular biology's representation of DNA.[2] Living with these shamans in the Amazonian rain forest, he saw repeated in their paintings two snakes wrapped around each other. He knew that the snake is an archetypal symbol found in most of the world's traditions and religions, and this strengthened his conviction that there is the link between DNA and serpent imagery. He concluded that with a little help from the hallucinogenic brew ayahuasca, shamans were able to enter into the consciousness or intelligence of their very cells and molecules and then paint what they had seen.

> *The serpent keeps recurring through the earliest cycles of mythology, always as a central symbol for the life of the universe and the continuity of creation. There are two great identical snakes on a Levantine libation vase of around 2000 BC, coiled around each other in a double helix, representing the original generation of life.*
>
> —LEWIS THOMAS, MD *The Lives of a Cell*

Inner Vision

Before I read Narby, I had already begun interpreting depictions of our biology in shamanic and sacred art. I hadn't been looking for symbolism when I began my adventures at the microscope. It was only after I had dwelt for some time in that micro-universe that I began to recognize its echoes in ancient art.

Until Narby's book came out, though, I thought the parallels that now seemed so obvious might be the result of a kind of dream—the product of a vivid imagination that no scientist would own up to. But I was excited by the connections I had begun to see everywhere. Perhaps they existed because, as some scientists maintain, God is hardwired into our brains; the need for spiritual connection is part of our nervous system. Seeing our biology reflected in art through the ages affirmed my belief that human consciousness has always understood that we are sacred beings who are divinely designed. And this invited me to ask, "What are we innately capable of knowing and seeing, unaided by scientific tools, theorems, and formulas?"

In ancient traditions, people used art, dance, and gesture to enfold and express sacred wisdom. They developed symbols and rituals to teach, honor, and touch the gods of creation. Mystical traditions developed meditations, visual constructions, and sacred art to guide people toward contact with the divine essence of life.

The great mythologist Joseph Campbell brought together a synthesis of science, myth, and the sacred. He wrote that scientific discoveries enabled us to recognize in the universe a reflection of our own inward nature, reconnecting us to ancient wisdom. Furthermore, Campbell said that we are leaping toward knowledge not only of our outer nature but also of our own deep inner mystery.[3]

It would not be too much to say that myth is the secret opening through which the inexhaustible energies of the cosmos pour into human cultural manifestation.

—JOSEPH CAMPBELL

Balloon Lady Becomes a "Shamanic Seeker"

Let me pause in this discussion and take a moment to describe my own experience of sacred art and ritual, beginning with a return to my "balloon lady" days. When I first took off my lab coat to spend time with children in the hospital, I would draw with them. Sometimes I'd ask them to draw their illness or how they felt. They drew whatever they wanted, and we talked about their images. I wasn't there as a scientist or to study healing, just to give them a few minutes of comfort in an uncomfortable, stressful environment. Still, many times I witnessed that this simple act of drawing seemed to bring relief and a kind of peace.

Images have the power to initiate healing without the cooperation of the intellect.

—CARL JUNG

In my own healing journey I discovered that expressing my feelings through art or dance released something deep inside. It could be tears, a sigh of relief, or an "a-ha!" moment. Drawing and moving became my most powerful teachers. From a cellular perspective, the two practices are connected as a whole, and through them I could express grief or anger, shame or elation. These were feelings and places I couldn't reach through words or my intellect. Healing, for me as for many others, can be attained through physical expression—through manifesting from within.

Art helped me maintain sanity during chaos; it brought me joy or deep tranquility simply in the doing. When I painted, I was in the zone; when I danced my sorrow or happiness, I was uplifted; when I walked in nature taking photographs in the vineyards, I was transported. The creative process took me someplace else—a place and state necessary for my well-being and healing.

For a period of time after I left the University of California School of Medicine, I worked individually with adults who had cancer using imagery as a healing tool. People came to me to learn how to meditate, manage stress, and use their imaginations for healing. They asked me to "conjure up" a visualization script for them. These were the days when guided visualization and imagery were just finding their way into supportive medical care, and I discovered that, somehow, I had a knack for receiving information from beyond the intellect and the skill to deliver custom guided visualizations.

If we were an indigenous culture, this work would be considered "shamanic," meaning that I accessed information by entering into an altered state of consciousness. Shamanic vision is an ancient gift from our human ancestors that I was able to tap into. In the clinic office, before the client arrived, I would alter my state of mind through meditation. I went even deeper into relaxation when the client and I were together in the room, and I led the client into a meditative state as well, to create conditions more conducive to healing. I knew that scientific research shows that slowing our brain waves down to an alpha or theta state facilitates hypnagogic imagery. In fact, visualization itself requires being in a deeply relaxed state—we have no choice but to relax, just as the indigenous shamans do, if we are to permit images to flow. When we are in a state of contemplation, deep relaxation, or prayer, our brain waves slow down to 4 to 10 cycles per second. Our normal awake state operates at beta brain waves, which range from 12 to 38 cycles per second.[4]

My goal in this work was to take a "shamanic journey" and "bring back" images for people to use on their own. Almost universally, the images resonated with my clients in a positive way. Still, I always questioned where the images came from. Was I reading minds? Had I tapped into "cosmic consciousness"? Was God speaking to me? Was I making it all up? Was I faking it?

Ultimately, the fact that I really didn't know the source of this imagery left me feeling insecure, and I stopped offering this service. In retrospect, I regret this—the work was genuinely helpful and useful. But my science mind ruled the day; my intellect was too uncomfortable operating in the intuitive realm of the great Mystery.

Shamanism: Our Oldest Healing Strategy

Since the beginning of recorded time, people have used pictures, symbols, and visualization as healing tools. This old, enduring system for healing with the imagination is called *shamanism.*[5] I am no shaman, but I've spent decades exploring the inner worlds and have apprenticed to a shaman. The word *shaman,* from the Siberian Tungus, refers to someone who can intentionally penetrate and translate the landscape of the imagination. They can alter their consciousness at will to enter what is often called "non-ordinary reality." Doorways to this mysterious realm are opened through drumming, chanting, dancing, psychoactive plants, meditation, dreams, and deep relaxation.

Through ritual the shaman reaches into the spirit realm in search of information that will help an individual or the tribe. This information often comes in the form of images, symbols, and songs, which the shaman interprets.

> *Shamanic traditions state that images, metaphors and stories are the best way to transmit knowledge. Myths are "scientific narratives" or stories about knowledge. Science means to know. . . . Wisdom requires not only the investigation of many things but contemplation of the mystery.*
>
> —JEREMY NARBY *The Cosmic Serpent*

Sharing Inner Knowledge: Story, Myth, Art, and Symbol

Let's widen the lens again and consider the role that stories, myth, and art play in passing along knowledge of human history. Through symbolic images, sculptures, and myths that have survived the people who created them, we have learned a lot about our ancient ancestors. Archeologists dig deep into our human past, learning the ancient beliefs, myths, and rituals of a culture from its art, temples, and writings.

Psychiatrist Carl Jung proposed that we carry within us unconscious images and myths that arise from our biological past, that our ancient memories have roots in our biological nature. He named this ancient knowing the *collective unconscious* and envisioned it as biologically inherited bits of experience. The collective unconscious retains and transmits the common psychological inheritance of humankind. Myths, legends, and symbols seen across time and place show the same forms as sacred and mythical. Such universal symbols are held in the collective unconscious and are called archetypal; they are primordial images or archaic remnants from our hidden past. According to Jung, as evolution of the embryonic body repeats its prehistory, so the mind also develops through a series of prehistoric stages. *According to Jung, archetypes are biologically grounded.*[6]

Jung believed that myths are inherited memories of the race that embody evolutionary processes and past experiences. And somehow, precognitive memories of these stories are carried within us, and we have access to this collective, unconscious stash of sacred signs and symbols. He proposed that we are all born with extensive foreknowledge and memories of the world. If this is so, where are those memories stored? In our DNA?

In *Art and Physics,* Leonard Shlain reminds us that the DNA molecule is a massive library of blueprints for everything from fingerprints to hair color and all the proteins in the body, yet not all of it is useful. In fact, as we learned in chapter 6, cell biologists revealed that the majority of our long strands of DNA provide no known information for any physical trait or molecule and called it "junk." Shlain asks whether this junk DNA with no present discernable value could be a source of our ancient memories. "It is not inconceivable that somewhere along its twisted, elongated shelves is a section from our evolutionary history."[7]

Perhaps this "silent DNA" serves as the depository of ancient memories. What if Shlain is correct that the bulk of our undecipherable DNA holds keys to healing wisdom and ancient knowledge? Does the almost-universal use of spiral movements in ancient sacred body practices tap into the very essence of our spiral molecules, our DNA, loosening memories from our collective unconscious vault? What if by engaging in tai chi, qigong, dance, and yogic kundalini movements, we unwind and rewind the cosmic karmic memories held in the twists and kinks of our long strands of DNA? Not only can moving our bodies help release present memories and change our gene expression, it may also give us access to ancient wisdoms.

> *When we visualize symbols they can have profoundly transformative effects. Symbols connect us with regions of our being which are completely unavailable to our analytical mind . . . they train us to understand . . . directly, jumping to . . . a deeper kind of understanding, which awakens intuition. A symbol can be a true reservoir of revelation.*
>
> —PIERO FERRUCCI *What We May Be*

The Spiral

The spiraling DNA was the first molecule I thought of as mythic and mystic—a symbol. In myth, the spiral symbolizes growth and transformation. It connects our roots in nature and our inner nature. Likewise, the spiral was one of the earliest recognized sacred symbols for many cultures.[8] Wherever it was found in nature, it was worshipped: as early as Paleolithic times, the whorled shell was revered. The curved and undulating spiral form was often included in images and sculptures of the oldest fertility symbols, the life-giving mother goddess. Did the DNA that guides our lives play a role in imparting to us symbolically the regenerative power of the spiral?

In nature, the spiral is characteristic of flow and growth. Galaxies grow through the inward spiraling of interstellar gases (see figure 8.1). The curve of an elephant's tusks, the pinecone, and the grapevine are all examples of spiraling growth, as is the nautilus shell (see figure 8.2). All are programmed, encoded by DNA that is itself a spiral. The movements of wind and water also follow this universal spiral pattern.

Figure 8.1 Spiral Galaxy M81

Spirals also have been associated with the passing on of the soul and have been used to decorate ancient burial chambers. The spiraling vine is a sacred Hebrew symbol for eternal life and is even represented in the ritual bread of the Jewish Sabbath, the challah.

If one is to understand the invisible, look carefully at the visible.

—THE TALMUD

The spiral is a form of transformation and beauty. Not only did it have a mythic presence in ancient cultures, but even today our biotechnology culture worships the sacred genetic spiral code as containing the answer to all life's ills.

Like all existence on the descending scale of realities, the spiral is a symbol. It denotes eternity, since it may go on forever. . . . This order, reverberating down into the microscopic and subatomic levels, both structures and reflects our consciousness

—JILL PURCE *The Mystic Spiral*

Figure 8.2 Many things grow in spirals, like plants, some seashells, and galaxies.

The Cell's "Code of Three"

DNA holds more than the spiral at its core physical and symbolic self: what about its coding abilities? Another element of cellular life that has reverberated throughout biology and taken root in our myths and symbols is the DNA "code of three." Life itself depends on threes: the three-letter genetic code necessary to create living things is a kind of Trinity. As we saw in chapter 6, of the four building blocks forming the genetic "scripture" of DNA, three of these structures at a time prescribe the correct amino acid to build our proteins. It is a three-letter code. Three states of cellular tension regulate gene expression, cell growth, life, and death: tight, loose, and just right. In chapter 4, we learned that cellular creation requires a 3^3 tube-like structure (the centriole) to guide the way.

Each of the two centrioles in a cell is made of microtubule pipes "welded" into nine groups of three each. Nine slightly twisted triplets link to form a hollow tube. Northwestern professor of molecular biology

Guenter Albrecht-Buehler asserts that this kind of universal design cannot be an accident of evolution; it must have arisen to serve a purpose.

Other design "coincidences" include the fact that at three weeks the cells of the human embryo enfold into three differentiated layers: ectoderm, mesoderm, and endoderm. And neither cork (from the cork oak tree) nor grapes for wine can be harvested until the plant has grown and matured for three years.

The triad is the form of completion of all things.
—NICOMACHUS OF GERASA Roman mathematician and philosopher

In biology, other essential triads include our triune brain, encompassing the brain stem, the limbic or emotional brain, and the cerebral cortex—the intellectual or thinking brain.[9] The reptilian brain stem provides for essential survival mechanisms; the limbic system gives us our emotional, nurturing capabilities; and the most recent addition, the outer cortex, enables us to think and reason.

Beyond our biology, there are other triplets. We who live on this third planet from the sun enjoy a life lived in three dimensions. In fact, astronomer and key figure of the seventeenth-century scientific revolution Johannes Kepler asserted that there are only three dimensions of space because of the holy Christian Trinity. In our storytelling, we naturally seem to love threes: we are offered three wishes, we have the three bears and three little pigs, and there were three magi at Jesus's birth.

The triangle is the world's preeminent symbol of divinity.
—MICHAEL SCHNEIDER *A Beginner's Guide to Constructing the Universe*

We are also drawn to the triangle. Three lines enclosing a space form the first stable physical structure, a symbol that is found in many religious traditions. It was an ancient Egyptian symbol of the Godhead; in Christianity, it is a sign of the Trinity. In Christian art, God's halo is traditionally a triangle; all others are round. It was a Pythagorean symbol for wisdom. It also represents the feminine (downward-pointing triangle) and the masculine (upward pointing).

In astrology, it came to symbolize water and fire, depending on which way the triangle points.

When we pray, we put our hands in the most natural pose and create a triangle. Hands in the prayer position also express the Buddhist greeting *namaste,* meaning "the divine in me greets the divine in you." When we complete a Taoist qigong energy practice, we form an upside-down triangle with our hands and place them on our belly to ground the energy. When we meditate, we may sit in a lotus position, forming a triangle with our entire body.

The Three of Spiritual Traditions

Christianity gives us the trinity—Father, Son, and Holy Spirit. Judaism has three faces of God—Yahveh or YHVH (the father), Shekinah (the feminine face of the divine), and Ruach (the breath of God). In the Jewish mystical tradition of Kabbalah, three columns of the Tree of Life depict the foundation of creation. The Hindu pantheon called the Trimurti includes Brahma (creator), Shiva (destroyer), and Vishnu (preserver or warrior).

> *Triads of gods appear very early, at the primitive level. . . .*
> *Arrangement in triads is an archetype in the history of religion,*
> *which in all probability formed the basis of the Christian Trinity.*
>
> —CARL JUNG *A Psychological Approach to the Dogma of the Trinity*

An ancient Greek creation myth tells that three immortal beings emerged out of chaos: Gaea, or Mother Earth; Eros, or Love; and Tartarus, the Underworld. A Navajo creation myth talks about three races: coyote, first man, and first woman. In addition to the Three Pure Ones, the Taoists have Three Kingdoms of the Universe (Heaven, Earth, and the Middle Kingdom of Man), while the Buddhists have Three Treasures or the Triple Jewel: Buddha, Dharma, and Sangha.[10] Table 8.1 lists sacred threesomes from a variety of traditions. Some of the triads represent three qualities of God or the holy spirit; others represent the pantheon of the three gods of creation.

Table 8.1 Sacred Triads of Spiritual Traditions

Spiritual Tradition	Triad		
Hindu	Brahma	Vishnu	Shiva
Ancient Greek	Gaea, Mother Earth	Eros, Love	Tartarus, Underworld
Hebrew	Yahveh, YHVH	Shekinah	Ruach
Christian	Father	Son	Holy Spirit
Navajo	First man	First woman	Coyote
Taoist	Heavenly jewel	Mystic jewel	Spiritual jewel
Buddhist	Buddha	Dharma	Sangha

And so we've seen that the essential pattern of three in biology is at the heart of creation—the genetic code, embryonic development, and the three domains of life. We embrace this pattern in our stories and myths, and we build it into how the Mysteries are structured and how our pantheons are peopled. As we weave this pattern through our cultures throughout time, are we expressing the wisdom inherent in our cells?

Three habits of the heart . . . the processes of connecting, nurturing, and integrating all of our cellular memories to create who we are, what we need, and what we have to give.

—PAUL PEARSALL, PHD *The Heart's Code*

Three in Your Life: Self-Creation

Why have I taken you on this metaphysical excursion? Because it links science, the sacred, and life itself, and I want to invite you to explore this pattern further. So many creation philosophies from ancient to modern times revolve around this powerful three—what might happen if we engaged the power of three to recreate our personal lives? Can we create positive change more easily by using our "three nature" to support a creative intent or emotional transformation? What might happen if we added to our meditation or movement practices the quality of three?

Exploration

A Triad of Intent: Discovering Your Own Power of Three

Become a mystic explorer, a cytonaut, by engaging the secrets of your cells. Experience for yourself whether creative change can be achieved more easily if you use a triad of intent. For example: will doing an exercise three times or saying a prayer three times have a generative effect? Will this repetitive pattern support change? Do it for three days, then three weeks, and record and watch what happens. Both our cells and sacred traditions indicate that *three is the key to making things happen.* Imagine the possibilities if you shift your practice in this way.

Body Prayer

A Cellular Creative Practice

Set aside some time to reflect on what you would like to change in your life or your world. Keep it simple and definable; write it down as an intention or a prayer. Now put that intention into a series of three movements or gestures. The following is one example.

I stand feeling my feet rooted to the earth. I am anchored in the earth. Reaching up with both hands open over my head, I stretch and reach toward the sky and voice my intent (silently or aloud): "I am grateful and open to receiving wisdom and providing it to others." Bringing my hands down in front of my heart in prayer position, I remind myself to honor the divinity in me, in others, and in all that is. I dedicate this body prayer to positive change in the world. Bending down, I touch my palms to the earth, planting my seed of intent, stating that I will do whatever it takes to achieve this. I repeat this three times.

Attach: Feel the earth under your feet. Be aware of the connection.

Align: Connect breath, ideas, feelings, and body. Consider doing some spiral movements and humming while tuning in to your intent.

Act: Plant the seed of intent. Listen and take action.

◇◇

It's as simple as one, two, three. Three brings a process to completion.

> *Without realizing it we "pierce polarity" whenever we count "one, two, three.". . . [This] reflects a major leap of consciousness. It gives us the ability to transcend polar bonds and realize the unlimited.*
>
> —MICHAEL SCHNEIDER *A Beginner's Guide to Constructing the Universe*

Cellular Wisdom in Sacred Art

Let's look at how the triangle and pattern of threes are repeated in sacred art. Here is the Hindu Sri Yantra used in meditation, which is said to symbolize the entire cosmos or the womb of creation (see figure 8.3). The triangles pointing downward represent the dynamic female principle of energy (Shakti). Those facing upward represent the static male principle of wisdom (Shiva). Constructed from nine interwoven triangles and a few surrounding circles, this "cosmogram" symbolizes sacred space for all the Hindu deities. At the center of the smallest circle is a dot known as the bindu, the point at which creation begins and unity becomes the many.

This was one of the first ancient mandalas that I saw as an apt symbol for a cell; the architecture seemed remarkably analogous. There is a center filled with triangular threesomes (DNA code?) and an outer circular rim like a cell membrane with "receptors" at the edge. Symbolically, for those who meditate on it, each of the features provides information and a place for focused intent.

Now take another look at Buckminster Fuller's geodesic dome (see figure 4.2 on page 75). You can see that triangles are key to its stable structure, just as they are in the Sri Yantra—just as in the fabric of our cells.

Figure 8.3 Hindu Sri Yantra

We know that everything is connected—our cells and molecules; mind, body, and spirit; modern science and ancient wisdom—in a grand design. Albert Einstein said, "We see the universe marvelously arranged and obeying certain laws but only dimly understand these laws." I've often thought this meant that he saw God as the generator of design. All life forms share the same secret codes and mysterious magic. We are sacred through and through, all the way down to our molecules. In the evolution of our new science, emphasis on quantum physics, energy, and the sacred has taken center stage. Yet consider that the formless energy of the quantum had to take form in our cells and molecules—and what divine forms, signs, and symbols we ourselves have become.

In closing this chapter, I will leave you with two symbols to reflect on, one ancient and one modern. The ancient symbol that first sparked my notion that the wisdom of the cells was hidden in plain sight in ancient structures and art was the medicine wheel I mentioned in the preface.

I saw this pictograph as representing a cell as well as the whole of life. Later I interpreted the three lines placed at each of the four directions as representing our centrioles, the magic triplets that direct our cells and thus our lives (see plate 4 in color insert).

The second symbol, a product of science and modern technology, was the other image that piqued my imagination early on. This is a computer graphic of the atomic structure of DNA we first saw in the preface (see plate 5 in color insert). It was created by professor emeritus Robert Langridge who was founder and director of the UCSF Computer Graphics Laboratory. Most people who see it think it's a painting of a mandala. Did this basic structure provide the inner inspiration for mandalas? People long ago often used circular mandalas to help them focus their minds and connect with the divine. Looking at this image could certainly do the same.

> *Then the people looked beyond their dwellings and fires and saw something infinitely more than the tangible world. They saw that the spiritual realm is the crystal mirror of the cosmos. What is seen reflects the essence of things not visible in the tangible world. . . . Then at times I enter the spiritual realm at its source. . . . And my own spirit soars!*
>
> —ANNA LEE WALTERS *The Spirit of Native America*

In this chapter I have offered the provocative idea that sacred art emerges from a direct knowing of the divine design of life. I have come to believe that it does. Also in the chapter, I have invited you to test the cells' "code of three" in your own life—to experiment with harnessing nature's creative power for your own growth and transformation.

In the next chapter we will conclude our journey by revisiting the winding trail we have traveled together as we have gained an intimate knowledge of the tiny sanctuaries of life that are common to us all.

Chapter 9

Connection–Cell-ebrate

One cannot but be in awe when [one]
contemplates the mysteries of eternity,
of life, of the marvelous structure of reality.
It is enough if one tries to merely comprehend a little
of this mystery every day. Never lose a holy curiosity.

—ALBERT EINSTEIN

Throughout this book we have used Nobel Prize–winning biochemist Christian de Duve's term *cytonaut*—sailor of the cell—to denote those of us who are willing explorers of the cell.[1] We have ventured inside the cell and discovered some of its biochemical mysteries. We also veered away from the microscope to gain a wider view of life, investigating the larger lessons our cells hold for us. I hope it has been an exhilarating adventure for you. Here we'll revisit some of our significant cellular explorations, to remember and "cell-ebrate" the marvels contained within this small, sacred vessel.

Interconnection

The cell, our oldest living ancestor, is the common ancestor for *all* life. Besides sharing the same DNA coding system with all other living creatures and plants, we all use the same elemental chemicals of carbon, hydrogen, oxygen, nitrogen, sulfur, and phosphorus as the basis for life. Our dynamic biochemical processes are also similar, getting their start in the tiniest and most ancient microorganisms.

I often have wondered: if we acknowledge this simple fact of shared "ingredients" and activities and accept it as the reality in which we all exist, will it help us recognize and respect our interconnectedness and the sanctity of all life? That has always been one of my goals for this book—to remind us of our connection with all others who share this planet. We are joined together by sharing the same elusive quantum physics, molecular DNA, and the essence of cellular life.

All living things require clean air, food, and water. Beautiful rain forests, all the other tree "people" in the world, and our growing farms and gardens generate the life-giving oxygen we breathe, while we exhale carbon dioxide, which plants transform into food—this is the ultimate recycling environment. What is one species' waste product (oxygen from plants, carbon dioxide from animals) is essential for the life of other species. It seems to me that this is a divinely designed partnership.

Unfortunately, many of us see no connection to people in remote areas of the planet, to distant forests or animals living in the tundra—or even to our next-door neighbors. We watch in horror as catastrophes unfold around the globe: the 2011 earthquake and tsunami in Japan and the threat of nuclear meltdown, the devastation wrought by tornados in the Midwest, the unspeakable suffering resulting from Hurricane Katrina, the appalling tragedy of starvation in Somalia. No matter how distant these events may be from us geographically, our cells and energetic fields are affected. Our very atoms travel through time and space, and it is said that we each may contain atoms that once spun in the bodies of Jesus and Buddha. It can come as no surprise that in the aftermath of the Japanese disaster, radioactive molecules traveled from Sendai, Japan, to New York City. We share the same air, water, world. Everything is connected, even if we can't see or feel the bond.

If what you have learned in this book helps support your life and nurture your cellular caretakers, will it also help you take better care of other creatures and their cells? What life-affirming contributions can we each make to the world we inhabit together? Will we begin to view our own cells and each other with greater reverence?

Original Blessing: Molecular Marriages

Our chemical universe that began about 4 billion years ago provides the foundation for all life. Amazing molecules carry information and the means for survival. The people who espouse intelligent design have at least one thing right—our molecules and cells carry intelligence. Intelligence is information. Molecular evolution is part of our heritage.

If molecules had never developed and found one another, there would be no life: perhaps God is a biochemist. Consider an idea we encountered early on—"molecular embrace." Dr. de Duve speaks of molecular complementarity, the embrace that is the very essence of how our molecules and biology work. Biological recognition is based on an essential, dynamic relationship between molecules. To engage in a cellular response, molecules must closely fit each other so that they can mold and bend to the other. Their bonding is an intimate exchange, one that is necessary for most molecular interactions of our cells. A truly cooperative endeavor is embedded into us at a microcosmic level so that life can thrive.

This universal phenomenon of molecular embrace is the essence of how we work at a chemical level—how our enzymes manage the chemical reactions in our cells. It's critical to immune recognition, information transfer, hormonal responses, drug reactions, and of course, DNA partnerships. Such a basic design in nature indicates how essential embrace and touch are to life—not only for cells and molecules but for us and for other creatures with whom we share the planet. Human infants will not survive without touch—something we learned the hard way decades ago. In hospital nurseries where orphans weren't held and nurtured, babies died. Only when people caring for them discovered that babies thrived with physical contact did we begin to learn how essential that simple gesture is to life. When mother cats, dogs, deer, and sheep lick their newborn babies, they stimulate the development of their babies' nervous systems. We need touch, all the way down to our atoms and molecules. As above (our whole selves), so below (our cellular lives): this ancient hermetic idea is revealed by all that we are.

Embrace Life

In chapter 1 we learned that our molecules embrace and merge with one another to create a sacred container for life's intricate machinations and the divine spark. They share or give away their electrons, another facet of cooperation at the molecular level.

We embrace our lovers, mothers, children, and friends. We embrace ideas that fit our values. We embrace the air that surrounds us so we can enjoy life from the inside out. When we embrace nature as an essential part of our existence, we transform who and what we are. We recognize that nature can be a sanctuary for us.

Only after I moved to a rural setting and joined a community garden did I begin to deeply experience nature. When we planted seeds in the earth just after dawn, I found the sacred in the natural world, the macrocosmic counterpart to the mysteries I had found under the microscope. The magical unfoldment of the life inside a tiny seed reawakened me to the awe I experienced while watching living cells. Recently I mentored school garden programs and was struck by how easily children can embrace nature when they learn to grow even a tiny bit of their food—one small carrot can do the job. There is wonder in such activity. We can honor our cells by reconnecting to nature. When we teach children where their food comes from, we help them grow their own roots and connections. With school gardens and in other innovative ways, we can demonstrate how the natural world is part of, not separate from, all of us.

Reflection

What do you embrace fully?

What or who embraces you or your ideas?

Who can you embrace right now?

Exploration

Embrace

Take a few minutes to tune in to your inner cytonaut and be aware that, working together, your trillions of cells hold and cherish you in an embrace.

Notice in your mind's eye that molecules and cells cooperate in this loving touch. They do not compete with one another to hold more of you; they share in creating the container of life for your spirit, self, and consciousness. Take a moment or two to experience gratitude for all they do.

If the only prayer you said in your whole life was, "Thank You," that would suffice.

—Meister Eckhart

Recognize I AM

Chapter 2, "I AM–Recognize," invites you to know that you are a sacred being. This is true even if religion is not part of your mind-set or way of life—your physical nature is itself sacred. Your body and mind, your molecules and cells, can be revered and held as your own exclusive expression of life. There will only be one of you, ever; you are unique.

We seem to more easily recognize our flaws and faults than what's right with us—the gifts we bring to the world, our laughter, our love, our creativity. Strive to see and accept *all* of who you are and have been—the one who was present on a road less traveled, the one who followed the crowd, the one who went down a path that shouldn't have been taken. There is a time to accept everything that has gone into crafting you as you are. Each decision has been your teacher and, hopefully, has brought you greater wisdom.

I recently taught a Cells and the Sacred workshop at a retreat center, an experience I touched on briefly in chapter 7, and I struggled. I had a hard time integrating my science self with my other, more sensitive spiritual side. I felt split in two. Before the final session, I went into the center's garden to gather myself together, to meditate. And it will come as no surprise to you by now that my cells carried me swiftly to *now.* I thanked them with all my heart for that gift, and as I did so, I heard an elusive voice say, *Finally, you got it—you really* mean *it!*

I had long made it a practice to thank my cells, and I thought I meant it every time, but this time I had the benefit of an altered state of consciousness: I was not remotely in my science mind, and it made a difference.

The biggest "aha!" moment was still to come. Hearing this message, *Finally, you got it,* filled me with a great giddiness and unleashed the voice further. It was as though I were suddenly receiving a transmission from my cells. This is what I heard:

> We are your oldest ancestors. We've survived millions of years, and yes, as you say, we certainly know how to live and thrive. The lessons you share describe what we must do as sacred beings. After all, we are holding the torch of life God gave us to carry. You carry a bit of that spark that resides in each one of us. Go to the place inside you that lets us be the spirit guides of your cells. Tap into us freely. Ask us how you can care for us, what you can do for us, and what we can do for you.

In a rush, my disparate parts came together—scientific knowing and spiritual seeking. I knew that what I had been teaching—which I had always thought of as my imagination extrapolating biological cellular life to a more poetic reflection of human life—was indeed a reflection of cellular reality. It was a felt sense of knowing, and it rang true. It was the voice of I AM THAT I AM.

Then the scientist in me stirred from silence and got busy, attempting to discount what I had just experienced. I was having none of it. If people through the ages have been able to "talk" to plants and get information from the visions that result, why can't our own cells provide information through visions? Why couldn't I have received the wisdom of our cells directly from my own cells all along? To a scientist who does not believe in the more holistic, feminine, and experiential side of science, this would no doubt sound like craziness or New Age "woo-woo." Yet the cytonauts who have journeyed through this book may be willing to accept this possibility. Our hunches, our inner voice, God, and intuition are all part of who we are. How—or whether—we express what comes from these inner spaces is a piece of the great adventure we call life.

Reflection

Recognize I AM THAT I AM.

Listen

In chapter 3 we learned how our cells listen to one another and how important it is to listen to others instead of reacting silently—arguing, judging, criticizing, and discounting. We also need to listen to our own inner voice, as I learned once again in the retreat center garden, especially when we are seeking guidance. Recall that our cells listen to all our chatter, receiving myriad molecular messages and deciding what action to take as a result. Many scientists say that only our neurons can interpret information, yet this is only part of the picture. All cells must be able to respond to incoming messages, and they do so at lightning speed, acting in nanoseconds—*billionths* of a second—and even picoseconds, *trillionths* of a second, well beyond the reach of our conscious awareness. Indeed, our cells apparently operate in quantum realities, each one carrying out a million maneuvers in time frames unfathomable to the human mind. They are in the *now* each moment: fluid, flexible, and ever changing.

When we stay in the moment and listen without reacting, our cells receive consistent messages and are able to choose wisely. They don't have to waste energy refereeing internal arguments or exciting one another needlessly. And this gives us the opportunity to thrive.

Reflection

Remember that your cells listen to everything happening in your world. When you are afraid, they coordinate their activities to help you deal with the situation, whether it is real or imagined. Your choice—to run away, freeze and hide, or change your mind about whatever challenge lies before you—changes your cells. Is the challenge worthy of getting your cells excited, or are you imagining danger where none exists?

Make it a practice to ask yourself: What messages are my cells listening to? Am I caring for them or making them work unnecessarily hard? Can I create a more nurturing partnership with them? Even if you look at cellular life lessons only as metaphor, the cells that carry you through life have much to teach you.

Strum the Strings of Life

Chapter 4 unlocked the secrets of intelligence held in our cells' architecture. Learning about the fabric and powerful strings of the cells marked a major turning point in my appreciation of the sacred nature of cells. Even though ancient healing strategies such as energy medicine, sound, and prayer are not yet accepted as part of mainstream medicine, these strategies are unwittingly based on the cells' architecture. Finally, we are able to identify anatomical structures inside our cells that can respond to these healing modalities: chanting, the flow of energy, the movement of breath.

Can we consider our cells shamans? Does the cell's very fabric shape-shift and move in its changed form for our higher good? Is any form of consciousness actually contained in those tubes and struts we can see under the microscope? My understanding of our cells' design has allowed me to see and appreciate the sacred origins of movement practices such as tai chi, yoga, dance, qigong—even walking. All of these stretch and soften the fabric within and around our cells. We may not be consciously aware of what our minds need to let go of for healing, but our cells are wise, and moving them can help healing unfold. Letting go is a central approach for both physical and spiritual healing. In fact, I propose that one of the most important practical lessons from our cells is the myriad ways we can let go—through movement, being in the present moment, humming, and writing down our truth.

Quantum physicists talk excitedly about a string theory of the universe, a concept well beyond my abilities to adequately describe or even understand.[2] But given what we have learned about the fabric of our cells, perhaps the strings of the universe and our cells' strings resonate when we are in a state of well-being, or perhaps when we are undergoing transformation or when the world faces a crisis. It is a provocative question to explore whether another way we and our cells are connected to all life on the planet is via strings. We do know that our cells' intelligence relies on its strings to act. Perhaps someday this will be added as another layer of string theory. As above, so below; as without, so within.

Reflection

What do I want to let go of?

What moves me, shape-shifts me to the higher expression of my life?

Who's pulling my strings?.

Sustain Energy

In chapter 5 we learned that our cells can only fully express themselves when they have the energy to do so. Our trillions of cells have a great capacity to churn out all the molecular energy we need to sustain us, yet what happens when or if our needs surpass their resources?

Consider that our cells generate about three pounds of ATP every day, which we use for seeing, breathing, pumping blood, moving, thinking, maintaining immune strength, replacing or repairing broken parts, and reading this book. How can we ensure that we don't run out of energy? What do we need to do—or not do—to balance the in-and-out flow of energy that powers our lives? We and our cells must rest and replenish, and we must invest our stock of energy wisely.

Reflection

Ensure you do not waste energy. Choose who you will spend time with, opting for the people who are most nurturing and stimulating in your life. Learn how to recognize who and what wastes your energy. Eliminate wasteful spending of your energetic resources.

Many years ago, a dear friend challenged by cancer told me that one of the first changes he had to make—knowing now that his energy was limited—was to say no to the people who were "energy sinks." Each of us, too, must answer these questions: How can we manage our energy resources more wisely, both our internal power supply and the energy we use in our external world? What kinds of food shall we choose? Who will be our friends? What will be our work and play?

I try to mirror my cells' wise use of energy in my daily life. I make sure I turn off lights in a room before I leave. I don't let water run when I'm brushing my teeth or washing dishes. I recycle and support local organic farmers as best I can; I purchase food, as I am able, that hasn't been raised with toxic chemicals. And I continue to learn what can sustain and build my energy in addition to the qigong practice I have engaged in for many years. Sitting and writing in my garden, amid fruit trees and singing birds, sustains the heart and soul of my energy. And sharing energy with others is another way to generate energy for myself. As we each pay attention to sustaining our cells and ourselves, using our resources wisely we help sustain our planet for future generations.

Reflection

Learn what sustains and builds your energy and mood.

Reduce stress by doing less.

Be cognizant of the footprints, carbon included, you leave behind.

Create a Legacy

In chapter 6 we explored genetic expression, and we left the chapter contemplating the expression of our life's purpose. I started writing this book about twenty years ago, and now I've come to a time when I've lived more years than I have left—an interesting perspective to have. As we enter our elder years, we naturally reflect on the paths we've taken and the blessings we've brought to the planet. A friend once advised, "Leave the place better for your having been here." My first legacy is love—my children and grandchildren. Through them I continue to learn how to share love.

Another legacy I leave is this book, a resource for you to be able to add to your knowledge as you shape and reshape your own legacy. If you share my belief that we are each here for a purpose, you will have discovered by now that your cells are here to help you fulfill it.

What is your purpose for being here? How will you leave the world a better place for your time here? What are you doing now toward that end, and what do you want to leave behind? We each have been granted the gifts of life energy and genes—how do we best nurture them for the highest good? For our own contentment and that of others?

Your genes—simple chemical codes—hold the patterns of information that built you. Shake out a packet of flower seeds or spinach seeds; these tiny containers hold all the information necessary to create the plant. Yet even with their influential genes, seeds can't do it on their own. They need tender care, water, and soil to manifest their majesty. The environment speaks to their genes just as the genes themselves communicate their instructions, and small changes in their environment can make a huge difference. I learned this lesson once again when I sprinkled beet seeds in two different containers. Though each container has the same soil and gets the same amount of sunlight, the beets are growing very differently because of the different shape and size of each container. One set has barely gotten started after weeks in its container, while the other waves its large leaves to all passersby. Just like seeds, your cells and self need nurturing and reinforcement in a healthy environment to manifest the best you.

Scientists are now showing us the possibility that our environment can influence which genes are expressed. So with the seeds. We can express our destiny, our legacies, our genes provided we are in the right environment. This certainly can be empowering when we appreciate that our genes are not the "boss of us."

REFLECTION

What is the one legacy you want to leave behind for present and future generations?

What can you do or be today that goes toward manifesting that purpose?

Learn and Remember

Chapter 7 explored creating cellular memory and habits. We learn by mentally and physically repeating an activity, whether it's writing, running, playing the piano, swimming, laughing, or loving. Each activity becomes wired into our neurons, patterned into our muscles, and held in the fabric of our cells. We can break old habits and make new ones, all with the help of our "sensory delight cells." We are hardwired for pain and pleasure; our survival depends on experiencing both, and our senses help us remember both. To learn new, positive behaviors and habits we can engage those senses that provide us with delight—the taste of a juicy apple, the scent of lavender or gardenia, a gentle touch on bare skin—all can help anchor the new lesson among our cells. Cellular memories depend on the roads more traveled. Cells that learn together create a trail to come back to more easily. Knowing that we can intentionally engage our cells in creating new memories and strengths, we have an enhanced sense of self-management. We can teach our cells and ourselves new tricks.

I have often struggled to stay with a regular exercise program. *Knowing* that it's good for me isn't enough. Once I get started, I may stick with it for a time, but until my cells like or even crave what I'm doing, it takes real commitment and discipline—neither of which is my strong suit. My belief—not yet proven, and perhaps you can join in the experiment—is that if we perform an action in cycles of three, we can imprint our cells and behavior and pave the way for success. "Three-ness" can help us create change. Our senses help us deliver and hold on to our intent and action.

An athletic friend visited on a recent weekend, and we took long walks both days. I took another long walk the following day—for a total of three days of consecutive walks. My mood improved, I had much more energy, and my focus was strong. My body-mind *liked* walking three days in a row, moving, breathing fresh air, and partaking of nature. Try your own sweet challenge for three days in a row. Then go for six, nine, and twenty-one days. Remember: it is said that it takes twenty-one days to change a habit; you can test that for yourself. After all, any practice you check out for yourself will reveal how *your* cells cooperate with *you.* My words here are not enough. Let your cytonaut self explore the rich potential for learning and knowing deep inside.

Each individual cell has amazing skills and fascinating architectural features, yet the collaboration and networking of these cells far exceed what each can do separately. The complexity and intricacy of our cellular universe surpasses the capabilities of the computers on our desks by a mile. Our cellular abilities are phenomenal, and the supportive networks they build allow the expression of our humanity. We have much to be grateful for, one cell at a time.

The senses are portals between our bodies and the world. It's where we meet and take the outside in.

—JAY MICHAELSON *God in Your Body*

Keep Wisdom

Our cells are both sanctuaries and wisdom keepers—the latter topic we explored in chapter 8. The delicate designs of our molecules form patterns as sophisticated and beautiful as a Rembrandt or Kandinsky painting, a Tibetan thanka (Buddhist religious paintings representing deities, cosmology, or mandalas), or a native medicine wheel.

When we create a home, a friendship, or an altar, it is a sacred space that our cells enjoy. We carry our feelings of sanctity within.

When we take time to really open our eyes, what designs do we see repeated in nature, in our own homes, and in the sanctuaries we visit? Can we consider that everything may contain or reveal wisdom? Is it possible that the forms and patterns revered for centuries may reflect the shapes our cells assume?

For a time, I led cancer support groups I called the Spiral Journey. On the first day we met, no one in the group recognized the spiral design as having any special meaning or relevance in their lives. Only when their eyes were opened to the myriad places this shape could be found did they become aware of its presence. Some of the participants were surprised and delighted to find that the art on their walls or the patterns on their sheets—items they had selected because they were pleasing or meaningful—contained spirals. Similarly, only when we look and *see* can we recognize the beauty, wisdom, and sacredness in each other.

REFLECTION

What carries and holds wisdom, within you and around you? Take some time to look around and notice the patterns in nature. Look at the leaves of trees, the petals of flowers, the shapes of rocks. Imagine the forms of your cells and their intricate designs. Watch a butterfly, a hummingbird, and a growing plant. Marvel at the designs we all share. What sacred forms do you resonate with?

Cell-ebrate

It is said that, in the end, how much we have worked, what books we've written, which scientific breakthroughs we've made are not important; what matters is how deeply we've loved and connected. I wrote this book to share the knowledge that we are loved by our powerful and magical cells, and I hope that you will pass this message along to others. We are never alone when we consider our cells are always with us, tiny containers of life reminding us of life's sanctity. Our cells are always in the now. Enjoy their great wisdom and rejoice with them—and then put your cells in motion. We can change the future and transform the present by moving—dancing, singing, sharing, progressing—by offering blessings, and by taking action.

Now, cytonaut that you are, you understand that you *are* your cells. You know that if you emulate their wise actions and operating instructions you can unburden yourself of needless mind chatter and instead be transported to enjoy all the deliciousness of life. My mind can give me all kinds of reasons *not* to act—and I know that you know what I'm talking about. Yet when I really think about my cells with compassion, my mind relaxes its grasp on *no, can't,* or *won't;* my cells take on a greater dimension and prominence in my well-being. When I consider the consequences of my habits for my cells, I make better choices. Doing so is not about "should"—for I truly want to care for my cells, vessels for the divine spark of God. Held in this light, making better choices is more about "can."

May you find yourself similarly moved to honor and nurture the sacred spark within you as you carry your new knowledge of the secrets of your cells forward into the future.

Now, I would like to share a blessing:

> Great spirit, mother father God, help me take better care of the trillions of sanctuaries that carry "me" in their hearts. Guide me to be aware of others, to see that we are all part of the sacredhood, that we are all connected, and that we each do our part to care, to love, and to ensure, as best we can, that tomorrow will be here for future generations—that we, as their ancestors, will respond well and leave nurturing legacies of wisdom. Amen.

As with all blessings, hold this one with intention and presence. Stop and *be* with your cells as a spiritual practice, once a day.

When I let myself become lost and wander in the matrix of my molecules and strings, I was embraced by the endless possibilities that we are quantum beings who exist in an energetic vibratory space uniquely occupied by all our energy, molecules, and cells. Our energetic configurations are unique; there truly is only one you in the whole of the universe. How lucky you are to have the role of *you* in this lifetime. How do you want to play that role?

Exploration

Lessons from Our Cells: Life's Operating Instructions, as Easy as 1, 2, 3

I leave you with one final exploration, to explore now and to take with you into the future. Make a copy of the following list of instructions you have received from your cells, place it on a wall you pass by often, and each day see which lesson might be your teacher. Remember that you can always return to these lessons to mine the mysteries and teachings of your cells.

Embrace sanctuary

Recognize self and other

Listen

Choose

Attach and let go

Sustain energy

Create purpose

Learn and remember

Keep and know wisdom

Connect and "cell-ebrate"

Acknowledgments

The people I want to acknowledge for guiding my way to this book encompass a lifetime of teachers, artists, scientists, colleagues, family, and friends. You provided me with insight and curiosity, and sometimes unwitting support and encouragement for my inner explorer. It has indeed taken a lifetime to reach this point. I have a deep and abiding gratitude to you all.

Thanks to the scientists upon whose shoulders I stand. I have reached into your work to find mine: my spiritual "grandfather," Albert Einstein, as well as Christian de Duve, Watson and Crick, Donald Ingber, and all the biochemists who dug deep into our cells to discover how they work.

And much gratitude to the bridge builders who brought together science and spirituality: to David Suzuki, PhD, who merges the knowledge of the First Nations and cell biology with the wisdom of nature and native knowing; to David Sobel, MD, and David Spiegel, MD, who showed how mind, social support, and our senses are part of the healing path; to Dean Ornish, MD, who put together ancient healing practices for modern medicine, and Mark Wexman, MD, who included me in a team engaged in the healing of hearts;

to Bruce Lipton, PhD, who looked at the cells as intelligent teachers and made cells friendly to nonscientists; and to Deepak Chopra, MD, who first took the cell into the deeper realms of consciousness and energy—to all of you I am indebted for the body of knowledge you have contributed to all of us.

To Larry Hershman, who brought me into the magical growing world of plants and the sanctuary of the garden; to Kristi Moya, who invited me into her life to learn about the healing power of shared meals; and to my parents, Natalie and Perk, who nurtured me with their love, food, and family and bequeathed Perk's writer genes to me.

Thanks to my teachers

To Max Rafelson, PhD, no longer earthbound, my science mentor and graduate advisor as I struggled in the lab to learn biochemistry at the University of Illinois College of Medicine. Thank you for helping me find my way as a researcher and learn to ask questions.

To Marshal Kadin, MD, now professor at Harvard University, who taught me how to use a microscope and look at the human cells that changed my life and perception of the world.

To the National Institutes of Health, the American Cancer Society, and the University of California for funding my medical research. You helped me discover I was actually a scientist.

To Anna Halprin, the first teacher who helped me learn that the body can express emotion and experience healing through movement, sound, and art.

To the aikido community of George Leonard, Richard Heckler, Richard Moon, and Wendy Palmer. You helped me get into my body in the first place, to become more than a "head."

To Tomas Pinkson, PhD, for decades of loving support and guidance in exploring other ways of knowing and teachings of indigenous people. Thank you for walking your talk and showing me the truth of being a heart warrior. To my friends at Wakan who shared the shamanic path with me in our journeys to become whole.

To Eli Jaxon Bear and Toni Varner (Gangaji), who on those early mornings in Bolinas taught me tai chi chi kung and the value of a daily practice in community.

To DaJin and Charlotte Sun, who opened me to the deep experience of qi and helped me know that there was, in fact, an invisible energy force.

And to Shirley Dockstader, who taught with me, helped me teach from the heart, and asked questions about our cells that made me think about them in ways I had not before. This book comes from our sharing knowledge and learning together.

Thanks to my dear, supportive, and loving friends

To Elson Haas, MD, whose Preventive Medical Center of Marin provided me with the first space in which to explore working with adult patients, and thereby to discover more of who I am. Your friendship, and the financial support by having me help you with your books and educational outreach, gave me the support I needed to write this book. I couldn't have done it without you.

To Mark Krigbaum, for our ongoing conversations and for being "called" to the work of our cells. You have produced incredible video and photography, and your encouragement and friendship have meant more than you will ever know.

To Jacki Fromme, Marilena Redfern, and Marcia Starck, whose years of friendship have sustained and nourished me, especially during times spent in isolation completing this book. I look forward to more celebrations and play. To the other Queen of Hearts, Beverley Kane, and to Ruben Kleiman, who shared many a "cell-ebration," editorial review, reflection, and wine-and-food adventure.

To Bethany Argisle, who has been a dear friend and a coach in getting the first round of this material done a decade ago, and who had me as a houseguest whether we wanted it or not. Words can't thank you enough for your continual encouragement and prayers, for making sure I was safe when I was scared, and for being my walking buddy. Thank you.

To John Harris, a muse and friend, whose heart connection pushed me to challenge my beliefs, live bigger and take risks, venture into emotional territories I didn't know existed, and articulate what is sacred.

To David Freed and Rabbi David White, who helped me reignite my Jewish roots and explore the mystical Kabbalah.

To all the people over the years who have come to my lectures, classes, and workshops: through your questions and being eager to learn, you are my greatest teachers. To all the people who have been part of my support groups, I thank you for opening my heart to being human and helping me glimpse what really goes on when someone is challenged by a life-threatening or life-diminishing illness. You have been the guides and benefactors of this work.

To Matthew Fox and Brian Swimme, who invited me to develop and teach this material early on. To all the other venues that welcomed me and my work, especially IONS and EarthRise Retreat Center, which ultimately brought Sounds True to me.

To Jennifer Y. Brown at Sounds True, the acquisitions editor who envisioned this as an influential book that will inspire people about science and the sacred. Thank you for helping me create a proposal that worked and for your ongoing support. To Haven Iverson, Tami Simon, and all the folks at Sounds True who made this book "sing." Thank you for your faith in me and this material.

To Sheridan McCarthy, my editor, who became more of a collaborator, fine-tuning and improving my words. I am so grateful for how you partnered with me, asking me questions, making me flesh out the missing pieces, adding your own words, and helping make this book so much more than I dreamed it could be.

To my children, Ted and Heather, who have questioned my assertions. You daily make me proud of how you've grown, raised your families, and brought your creative minds and hearts to your own work. Thank you for listening to me kvetch and try to articulate my "way out" ideas. Your love and presence make all my struggles worth it.

To my grandchildren—Ethan, Harper, Benny, and Micah—to you I offer a view of the world that, when you are old enough to understand it, may nurture, help you grow, and inspire you.

Of course, to my cells, for getting me here in reasonably good health and fostering me to complete this work even when I doubted—for showing me the way.

And to God Hashem, for it is with you I have had the biggest struggle and the greatest connection. It has been in my search for you that my cells have become holy and love has become the most essential ingredient of my life.

Appendix 1

Energy Mapping Graph

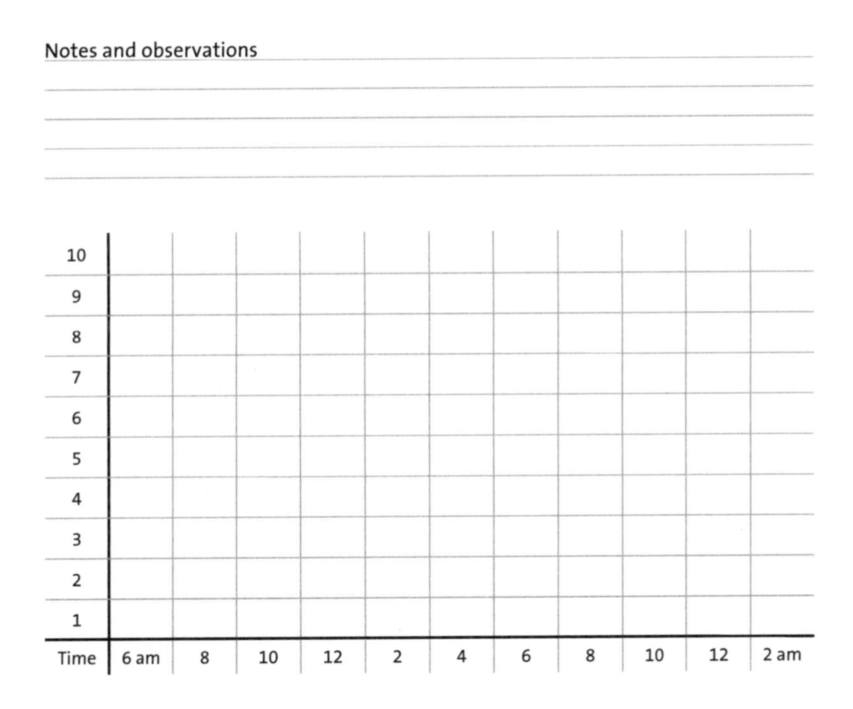

Map your energy rhythms: 10 the highest; 1 the lowest. Record at least 5x/day.

Energy ■ Mood ○ Tension ▲ Choose your own symbols or colors

Appendix 2

Qigong Body Prayer Series

I hope you have enjoyed engaging your cells with the qigong body prayers as you worked your way through the book. Here is a full series of qigong movements (which also appears in chapter 5). I want to point out, however, the value of doing even just one of the sequences. It is easy to learn and remember one at a time and to teach it to your children, friends, and family. Also, there may be only one or two that you feel moved to do at any given time. This is a time to really listen to what you need and sail ahead in the best direction. Your cells will love them all.

In preparation for doing any qigong practice, use your good sense and avoid doing it outside if it's too windy or too cold. Wear loose, comfortable clothing. When you practice inside, take off your shoes and feel the floor under your feet. You can do the same outside if the area is safe for your delicate soles and toes. Relax and be prepared for a cell treat.

The more you practice, the more you will be able to depend on the calm and nourishing energy you are cultivating. I recommend doing each sequence at least three times. The Taoists who originated these movements held three as a sacred number (imagine that!), so you can do them in series of threes. When I was taught many of these, we did each one eighteen or thirty-six times.

Give yourself at least ten to twenty minutes to do this practice. When I do the entire cycle as it is described here, I do a minimum of three repetitions of each form, except for Waist Circles; I do as many as I want of those. I often find myself automatically doing Waist Circles whenever I'm standing and waiting for something—in line at the movies, for instance. Try it.

The Basic Posture: Standing Home Alignment

Feel your feet on the earth, grounded and anchored. Place your feet shoulder-width apart, parallel to each other. You can imagine roots from the soles of your feet that reach deep into the earth. To help find that solid and centered place on your feet, rock back and forth and then sideways until you feel yourself grounded in the earth. You can gain strength from the earth's energy when you feel your feet upon her. You may also perceive or imagine that you are drawing up the earth's energy through your feet.

Your knees are slightly bent, your butt tucked under. Your shoulders are dropped and relaxed. Your arms hang loose at your sides. Your tongue rests softly on the roof of your mouth behind your teeth. (This is calledthe*innersmile*,andyoucanpracticeitanytime.)Yourchinisparalleltothe floor; you can imagine a golden cord connecting your head to heaven, a link to another source of energy.

Rock a bit until you feel solid on the ground.

All movements of this series start with taking this basic stance.

Another option of Standing Home is to assume this posture and then bend your elbows and place them at your sides by your waist. Hands are open and palms are facing each other at the level just below your belly button. This now becomes the Standing Stake, a standing meditation in which you begin to generate qi. Remember to keep your knees gently bent, and when you want to explore this, do it for a few minutes. With some teachers, this is the very first practice a student will be taught. They will work up to standing thirty minutes. It certainly strengthens your legs, body, and resolve.

Rooting and Spiraling: Waist Circles

This is similar to the practice included in chapter 2, and another variation of it appears in chapter 6. Starting from the basic posture, begin to inscribe circles with your waist, rotating belly and hips as though you are a rope held between the earth and sky, moving like you are using a Hula-Hoop, with your shoulders and chin remaining parallel to the earth. Continue spiraling your waist until you feel anchored, and then change direction. You may discover that one direction feels easier and more natural than the other. At this point, for a self-sustaining, simple practice, begin to hum. You can stop here or continue on.

Opening to Breath

Either stand in the basic posture or sit and place your hands in front of your belly (dan tien). Bring the backs of your hands together, palms facing out, in front of your belly with your fingertips pointing down. Elbows are bent and hands are relaxed. Relax your shoulders.

Gently push your hands out to the sides as if you are pushing open curtains. You are creating space in your belly for more breath or qi. Inhale as you do this.

When you are full of breath, turn your hands around with palms facing each other and fingertips facing away from the body. While exhaling, bring them together in front of your dan tien until they are about six inches apart.

Once more, place the backs of your hands together and push open the curtains, and then bring your hands back over your belly. Repeat this sequence at least one more time.

When you first begin this form, don't worry about your breathing; just breathe naturally. Your movement will follow your breath. The rhythm is to separate your hands outward as you inhale, and as you bring your hands back toward the center, gently exhale.

This is a particularly relaxing form, and a good mini-sequence to do is the basic posture, rooting, spiraling, humming, and opening to breath.

Energy Wash

This part of the sequence is perfect to do when you want to relieve the mind of unwelcome thoughts or stress.

Stand rooted with arms loose at your sides. Raise your arms at your sides, elbows slightly bent and palms facing up, fingertips pointing outward away from the body. Inhale while you are raising your arms until they are directly above your head. Palms now face each other, elbows softly bent. When hands are above your head, fingertips are gently curved, facing up toward the sky.

Pause and exhale while you imagine receiving qi from heaven or the universe.

When you are ready, inhale and turn your palms down, toward the crown of your head. Spread open your fingers and with your palms facing downward, slowly lower your hands in front of the midline of your body, imagining clear new qi flowing from your fingertips while the energy you don't need is being washed out. You might imagine that new energy is being sent to every cell. Take as long as you need to lower your hands while you "wash."

If you come to a place where you can't feel the energy or it feels dense, keep your hands there until you notice a change. And you may not feel anything at all.

When your hands reach your thighs, visualize the wash continuing down your legs and feet, and then shake out your hands, sending all the "used" energy back into the earth to be recycled.

You can repeat this sequence as frequently as you need. I recommend at least three times per session. It's especially helpful when you are feeling anxious or tired or have a lot of unwanted mind chatter. Right before an important meeting or presentation is another good time for an energy wash.

Sipping Qi

This is like a reversal of the Energy Wash.

Your hands are in front of your belly, palms cupped upward. Raise them slowly up the midline of your body to your lips.

With elbows bent, turn your hands so that they are now pushing up as you raise your arms above your head, palms facing upward until elbows straighten.

Now spread your arms extended out to your sides, palms facing down. Lower your arms down the sides of your body. Cup your hands at your belly and begin again. Repeat a minimum of two more times.

A combination of the Energy Wash and Sipping Qi is a good refresher sequence, removing what you don't need and filling with new energy. Always remember to be loose and relaxed while doing any of the series. Enjoy!

Core Wave

You may recognize this as a basic tai chi movement. It also can be done by itself for relaxation.

Start in the Standing Home posture. Remember to create the inner smile. Your arms are at your sides.

In a wavelike, easy movement, raise your arms slowly in front of you, about shoulder-width apart, wrists relaxed. Move your arms in wavelike rounded motions. Bring your hands and arms no higher than your chest. Your wrists stay soft, elbows are dropped and relaxed, and palms face downward.

Now lower your arms with palms facing down.

Allow your whole body to get into this movement, not simply your arms. You may feel as if you are pumping energy up your back and from your feet as you raise your arms. Experience yourself as flowing waves of water, fluid as the waters within you. Remember to breathe.

Heart Thymus Wave

This is a good tensegrity movement that stretches you and helps strengthen the thymus, one of the major immune organs and birthplace of our T cells. When you are done with this sequence, you can gently tap your chest with your fingertips above your sternum. You can also hum to your thymus. The thymus tap and hum is said to be a good preventive strategy during cold and flu season.

Continuing from the core wave, when your arms and hands reach chest height, extend them out to the sides with palms facing forward. Turn your head to one side and flex your wrists back.

Bring your palms toward each other in a soft, flowing wave, and then extend again, turning your head to the other side, once more flexing your wrists.

Bring your palms toward each other and lower your arms, palms down. Repeat two more times.

Integration: Balancing Yin and Yang, Right and Left Hemispheres

This is another tensegrity movement. It also balances the right and left hemispheres of the brain and is equivalent to alternate nostril breathing in yoga.

Beginning in the same Standing Home posture as all the other poses, bring your right hand in front of your belly, palm down. Your elbow is gently bent. Your left hand is hanging straight, not rigid, at your side, palm down.

Raise both your arms simultaneously. Extend the left arm out to your side while the right rises along the midline of your body. Continue until they both reach above your head, fully extended, palms facing one another. Pause.

Turn both palms down, with your left palm now going down the midline and the right arm extended out to the side. Slowly lower both arms.

Now reverse the sequence. When your arms reach the level of your belly, raise your left arm up the center while your right arm rises to the side. Repeat this three times on each side or until you get the rhythm of the movement.

This sequence took me *weeks* to learn, so be easy on yourself. When I recently taught this series, most in the class got it on the first try while one person never got it.

Tip: This is an exercise you have to let your body learn without your mind trying to figure it out.

Gathering and Storing the Qi: Closing the Circuits

When you are finished practicing qigong, you always gather in the qi and "close the circuits."

Take the Standing Home pose and cup your hands in front of your lower dan tien, your belly. Now widen your stance and reach behind and around you, gathering the qi in a circular embrace. Embrace this qi in front of your belly and then press your palms close to your body, forming an upside down V with your hands. Remain in this position for a few moments. This is another position in which you can simply stand, relaxed, with gently bent knees and inner smile. Close your eyes and let the qi move you, fill you, and replenish your cells. This can be another form of a standing meditation.

If you've never practiced tai chi or qigong, it's always useful to work with an experienced teacher. You may also find it worthwhile to keep a journal and occasionally map your energy and watch what happens. When Shirley Dockstader and I were developing this series for classes we were teaching at California Pacific Medical Center in San Francisco, her energy mapping showed a relatively steady and high energy level, whereas mine showed extreme peaks and valleys that balanced out somewhat the more regularly I did this practice. Shirley was a longtime practitioner of qigong, whereas I was a beginner at that time, almost twenty years ago.

To track your own energy level, make about a dozen copies of the template of the energy graph in appendix 1. Map your energy daily, noting if there's a consistent pattern of highs and lows. For a first go at this, I recommend paying attention to your qualities of energy for at least a week, until a pattern emerges or you observe a consistent low time of the day. Then you can take one of two approaches with the qigong practice. The first is to choose a time to practice the sequence daily for at least ten minutes. Map your energy to see if your pattern has changed or your energy has been raised after the qigong. The second approach is even simpler. Use an energy graph to evaluate your energy level both before and after your qigong practice. Mark the graph with your energy level before the exercises, practice the qigong series, and then assess your energy and mark the graph again. You can use one page simply for marking energy, mood, and tension before and after practice. It becomes very convincing if you "cultivate energy" at a low point of your day.

Notes

Chapter 1 Sanctuary–Embrace

1. Lauterwasser, *Water Sound Images,* 12, 38–42.
2. Hart and Stevens, *Drumming at the Edge of Magic,* 11.
3. Teilhard de Chardin, *The Phenomenon of Man,* 113.

Chapter 2 I AM–Recognize

1. Vincent and Revillard, "Characterization of Molecules Bearing HLA."
2. American Autoimmune Related Diseases Association, "The Cost Burden of Autoimmune Disease."
3. Rose, "Mechanisms of Autoimmunity."
4. Weinshenker, "Natural History of Multiple Sclerosis."
5. Macfarlane, "Olfaction in the Development of Social."
6. Wedekind, "MHC-Dependent Mate Preferences in Humans."
7. Laurance, "Why Women Can't Sniff."

8. German et al., "Olfaction, Where Nutrition, Memory."
9. Demarquay, Ryvlin, and Royet, "Olfaction and Neurological Diseases."
10. Lafreniere and Mann, "Anosmia: Loss of Smell."
11. Cheney, "Chronic Fatigue, Mycotoxins, Abnormal"; and Cheney, "New Insights into the Pathophysiology."
12. Reichlin, "Neuroendocrine-Immune Interactions."
13. Haffner, "The Metabolic Syndrome: Inflammation."
14. Stoll and Bendszus, "Inflammation and Atherosclerosis."
15. Rood et al., "The Effects of Stress and Relaxation."
16. Bartrop et al., "Depressed Lymphocyte Function."
17. Mahlberg, "Therapeutic Healing with Sound."
18. Dr. Angeles Arrien, personal communication in a course taught by Dr. Arrien.

Chapter 3 Receptivity–Listen

1. de Duve, *Vital Dust.*
2. Siegel et al., *Basic Neurochemistry.*
3. Stapleton, "Sir James Black and Propranolol."
4. Hassett, "The Sweat Gland."
5. Sapolsky, *Why Zebras Don't Get Ulcers.*
6. Kabat-Zinn, *Wherever You Go, There You Are.*
7. Pennebaker, Kiecolt-Glaser, and Glaser, "Disclosure of Traumas and Immune Function."
8. Smyth et al., "Effects of Writing About Stressful."
9. Berkman and Syme, "Social Networks, Host Resistance and Mortality."
10. Bruhn, "An Epidemiological Study of Myocardial Infarctions."

11. Astin et al., "Mind-Body Medicine."
12. Cohen, Tyrell, and Smith, "Psychological Stress and Susceptibility."
13. Heinrichs et al., "Social Support and Oxytocin Interact."
14. Taylor, *The Tending Instinct.*
15. Kroeger, "Oxytocin: Key Hormone."
16. Naber et al., "Intranasal Oxytocin Increases Fathers'."
17. Byrd, "Positive Therapeutic Effects."
18. Harris et al., "A Randomized, Controlled Trial."
19. Dossey, "The Return of Prayer."

Chapter 4 The Fabric of Life–Choose

1. Ingber, "The Architecture of Life."
2. Caspar, "Movement and Self-Control."
3. Ingber, "Cellular Tensegrity."
4. Fuller, "Tensegrity."
5. Castaneda, "Magical Passes."
6. Horgan, "Consciousness, Microtubules, and the Quantum."
7. Desai and Mitchison, "Microtubule Polymerization Dynamics."
8. Ron Nadeau, personal communication, Fort Bragg, CA.
9. Ainsworth, "Stretching the Imagination."
10. Paszek et al., "Tensional Homeostasis and the Malignant Phenotype."
11. Evans, "Substrate Stiffness Affects Early Differentiation."
12. Winkelman, *Shamanism.*
13. Castaneda, "Magical Passes."

14. Albrecht-Buehler, "Autonomous Movements of Cytoplasmic Fragments."
15. Albrecht-Buehler, "A Rudimentary Form of Cellular 'Vision.'"
16. Albrecht-Buehler, "Does the Geometric Design of Centrioles."
17. Penrose, *The Emperor's New Mind.*

Chapter 5 Energy–Sustain

1. Einstein, "Ist die Trägheit."
2. Cohen, *The Way of Qigong.*
3. Margulis and Sagan, *Microcosmos,* 31, 33, 128–136.
4. Palomaki et al., "Ubiquinone Supplementation During Lovastatin Treatment."
5. Moons, Eisenberger, and Taylor, "Anger and Fear Responses to Stress," 24, 215–19.
6. Thayer, "Energy, Tiredness, and Tension," 119.
7. Barrett, *Molecular Messages of the Heart.*
8. Wolf et al., "Reducing Frailty and Falls in Older Persons."
9. Sheldrake, *The Rebirth of Nature.*

Chapter 6 Purpose–Create

1. Barrett, "Induction of Differentiation Markers."
2. The GDB Human Genome Database Hosted by RTI International [online], North Carolina, gdbreports/CountGeneByChromosome.html,
3. Elgar and Vavouri, "Tuning in to the Signals."
4. "Genes and Chromosomes," Centre for Genetics Education. Internet: genetics.edu.au.

5. If you are intrigued by numerology, see Angeles Arrien, *The Tarot Handbook: Practial Applications of Ancient Visual Symbols* (New York: Tarcher/Putnam, 1997.)
6. Misteli and Spector, eds., *The Nucleus.*
7. Crick, "The Genetic Code."
8. Watters, "DNA Is Not Destiny."
9. Waterland and Jirtle, "Transposable Elements."
10. Lipton, *The Biology of Belief.*
11. Li and Ho, "p53-Dependent DNA Repair and Apoptosis."
12. Hardy, "Apoptosis in the Human Embryo."
13. Eisenberg, "An Evolutionary Review of Human Telomere Biology."
14. Vogelstein and Kinzler, "The Multistep Nature of Cancer."
15. Kadouri et al., "Cancer Risks in Carriers."
16. Bennett, "Molecular Epidemiology of Human Cancer Risk."
17. Selivanova, "p53: Fighting Cancer."
18. Pfeifer et al., "Tobacco Smoke Carcinogens."
19. Ming et al., "Stress-Reducing Practice of Qigong."
20. Oh et al., "A Critical Review of the Effects."
21. Syrjala et al., "Relaxation and Imagery and Cognitive-Behavioral Training."

Chapter 7 Memory–Learn

1. Barrett, *Molecular Messengers of the Heart.*
2. Bartolomeo, "The Relationship between Visual Perception and Visual Mental Imagery."
3. Childre and Martin, *The Heartmath Solution.*
4. Pribram, *Languages of the Brain.*

5. Rossi, *The Psychobiology of Mind-Body Healing.*
6. You can condition Pavlov's dog online; see "Pavlov's Dog," Nobelprize.org, August 14, 2011, nobelprize.org/educational/medicine/pavlov/.
7. Ader and Cohen, "Behaviorally Conditioned Immunosuppression."
8. Ader, "Conditioned Immunopharmocological Effects in Animals."
9. Barrett, "Psychoneuroimmunology: Bridge between Science and Spirit."
10. Slagter, "Mental Training as a Tool."
11. Rosen, *My Voice Will Go with You.*
12. Sheikh, *Imagination and Healing.*

Chapter 8 Wisdom Keepers–Reflect

1. Shlain, *Art and Physics.*
2. Narby, *The Cosmic Serpent.*
3. Campbell and Moyers, *The Power of Myth.*
4. Fell, Axmacher, and Haupt, "From Alpha to Gamma."
5. Harner, *The Way of the Shaman.*
6. Jung, *Man and His Symbols.*
7. Shlain, *Art and Physics,* 413–414.
8. Purce, *The Mystic Spiral.*
9. MacLean, "The Triune Brain in Conflict."
10. Beliefnet.com, "The Three Jewels of Buddhism."

Chapter 9 Connection–Cell-ebrate

1. de Duve, *Vital Dust.*
2. Capra, *The Hidden Connections.*

References

Aaron, Rabbi David. *The God-Powered Life: Awakening to Your Divine Purpose.* Boston: Trumpeter, 2009.

Achterberg, Jeanne. *Imagery in Healing: Shamanism and Modern Medicine.* Boston: Shambhala, 2002.

Achterberg, Jeanne, Barbara Dossey, and Leslie Kolkmeier. *Rituals of Healing: Using Imagery for Health and Wellness.* New York: Bantam, 1994.

Ader, R. "Conditioned Immunopharmocological Effects in Animals: Implications for Conditioning Model of Pharmacotherapy." In *Placebo: Theory, Research and Mechanisms,* edited by L. White, B. Tursky, and G. Schwartz, 306–23. New York: Guilford Press, 1985.

———. *Psychoneuroimmunology.* New York: Academic Press, 1991.

Ader, R., and N. Cohen, eds. "Behaviorally Conditioned Immunosuppression." *Psychosomatic Medicine* 37 (1975): 333–40.

Ader, R., D. Felten, and N. Cohen. *Psychoneuroimmunology.* New York: Academic Press, 2001.

Ainsworth, Claire. "Stretching the Imagination." *Nature* 456 (December 2008): 696–99.

Albrecht-Buehler, G. "Autonomous Movements of Cytoplasmic Fragments." *Proceedings of the National Academy of Sciences of the United States of America* 77 (1980): 6639–43.

———. "Cell Intelligence." basic.northwestern.edu/g-buehler/FRAME.HTM.

———. "The Cellular Infrared Detector Appears to Be Contained in the Centrosome." *Cell Motility and the Cytoskeleton* 27 (1994): 262–71.

———. "Changes of Cell Behavior by Near-Infrared Signals." *Cell Motility and the Cytoskeleton* 32 (1995): 299–304.

———. "Does the Geometric Design of Centrioles Imply Their Function?" *Cell Motility* 1 (1981): 237–65.

———. "Role of Cortical Tension in Fibroblast Shape and Movement." *Cell Motility and the Cytoskeleton* 7 (1987): 54–67.

———. "A Rudimentary Form of Cellular 'Vision.'" *Proceedings of the National Academy of Sciences of the United States of America* 89 (1992): 8288–92.

American Autoimmune Related Diseases Association (AARDA) and National Coalition of Autoimmune Patient Groups (NCAPG). 2011 Study Report: "The Cost Burden of Autoimmune Disease."

Amundson, S. A., T. G. Myers, and A. J. Fornace Jr. "Roles for p53 in Growth Arrest and Apoptosis: Putting on the Brakes after Genotoxic Stress." *Oncogene* 17, no. 25 (December 1998): 3287–99.

Apanius, V., et al. "The Nature of Selection on the Major Histocompatibility Complex." *Critical Reviews in Immunology* 17 (1997): 179–224.

Argüelles, José. *The Mayan Factor.* Santa Fe, NM: Bear & Company, 1987.

Arrien, Angeles. *The Tarot Handbook: Practial Applications of Ancient Visual Symbols.* New York: Tarcher/Putnam, 1997.

Astin, J. A., et al. "Mind-Body Medicine: State of the Science, Implications for Practice." *Journal of the American Board of Family Practice* 16 (2003): 131–47.

Aubert, G., and P. M. Lansdorp. "Telomeres and Aging." *Physiological Reviews* 88, no. 2 (April 2008): 557–79.

Barrett, S. "Induction of Differentiation Markers on Human Acute Leukemia Cells." *Blood* 51 (1978): 625a.

———. *Molecular Messengers of the Heart.* KABA, 2002. Compact disc.

———. "Psychoneuroimmunology: Bridge between Science and Spirit." In *Silver Threads: Twenty-Five Years of Parapsychology Research*, edited by B. Kane, J. Millay, and D. Brown, 170–80. New York: Praeger, 1993.

———. "Psychoneuroimmunology: Bridge between Science and Spirit." In *Radiant Minds: Scientists Explore the Dimensions of Consciousness,* edited by Jean Millay, 65–79. Doyle, CA: Millay, 2010.

Bartolomeo, P. "The Relationship between Visual Perception and Visual Mental Imagery: A Reappraisal of the Neuropsychological Evidence." *Cortex* 38 (2002): 357–78.

Bartrop, R. W., et al. "Depressed Lymphocyte Function after Bereavement." *Lancet* 1 (1977): 834–6.

Beliefnet.com. "The Three Jewels of Buddhism." Excerpted from Robert Thurman, *The Jewel Tree of Tibet* (New York: Free Press,

2005). beliefnet.com/Faiths/Buddhism/2005/04/The-Three-Jewels-Of-Buddhism.aspx#ixzz1VDhHWS9r.

Bennett, Mary Payne, and Cecile A. Lengache. "Humor and Laughter May Influence Health: I. History and Background." *Evidence-Based Complementary and Alternative Medicine* 3, no. 1 (March 2006): 61–63.

Bennett, William P. "Molecular Epidemiology of Human Cancer Risk: Gene–Environment Interactions and p53 Mutation Spectrum in Human Lung Cancer." Special issue, *Journal of Pathology* 187, no. 1 (January 1999): 8–18.

Benson, H., with Marg Stark. *Timeless Healing: The Power and Biology of Belief.* New York: Simon & Schuster, 1996.

Benson, H., et al. "The Relaxation Response." *Psychiatry* 37 (1974): 3746.

Berk, L., et al. "Humor Associated Laughter Decreases Cortisol and Increases Spontaneous Lymphocyte Blastogenesis." *Clinical Research* 36 (1988): 435A.

Berk, L., et al. "Eustress of Mirthful Laughter Modifies Natural Killer Cell Activity." *Clinical Research* 37 (1989): 115A.

Berk, L., et al. "Modulation of Neuroimmune Parameters During the Eustress of Humor-Associated Mirthful Laughter." *Alternative Therapies in Health and Medicine* (2001): 62–72, 74–6.

Berkman, L. F., and S. I. Syme. "Social Networks, Host Resistance and Mortality: A Nine-Year Follow-up Study of Alameda County Residents." *American Journal of Epidemiology* 109 (1979): 186–204.

Besedovsky, H. O., et al. "Hypothalamic Changes during the Immune Response." *European Journal of Immunology* 7 (1977): 323–25.

Biémont, C., and C. Vieira. "Genetics: Junk DNA as an Evolutionary Force." *Nature* 443, no. 7111 (2006): 521–24.

Bishop, J. M. "The Molecular Genetics of Cancer." *Science* 235, no. 4786 (January 1987): 305–11.

Boorstin, Daniel J. *The Discoverers: A History of Man's Search to Know His World and Himself.* New York: Random House, 1983.

Braun, W. E. "HLA Molecules in Autoimmune Diseases." *Clinical Biochemistry* 25 (1992): 187–91.

Brill, et al. "The Role of Apoptosis in Normal and Abnormal Embryonic Development." *Journal of Assisted Reproduction and Genetics* 16, no. 10 (1999): 512–19.

Bruhn, J. G. "An Epidemiological Study of Myocardial Infarctions in an Italian-American Community." *Journal of Chronic Diseases* 18 (1965): 353–65.

Bulloch, K. "Neuroanatomy of Lymphoid Tissues: A Review." In *Neural Modulation of Immunity,* edited by R. Guillemin et al., 49–85. New York: Raven Press, 1985.

Butcher, Darci T., Tamara Alliston, and Valerie M. Weaver. "A Tense Situation: Forcing Tumour Progression." *Nature Reviews Cancer* 9 (February 2009): 108–22.

Byrd, R. C. "Positive Therapeutic Effects of Intercessory Prayer in a Coronary Care Unit Population." *Southern Medical Journal* 81, no. 7 (1988): 826–9.

Campbell, Joseph, and Bill Moyers. *The Power of Myth.* Edited by Betty Sue Flowers. New York: Doubleday, 1988.

Campeau, P. M., et al. "Hereditary Breast Cancer: New Genetic Developments, New Therapeutic Avenues." *Human Genetics* 124, no. 1 (2008): 31–42.

Capra, Fritjof. *The Hidden Connections: Integrating the Biological, Cognitive, and Social Dimensions of Life into a Science of Sustainability.* New York: Doubleday, 2002.

Caspar, Donald. "Movement and Self-Control in Protein Assembly." *Biophysical Journal* 32 (October 1980): 103–38.

Castaneda, Carlos. "Magical Passes." *Yoga Journal* (January/February 1998): 74–84.

———. *Magical Passes: The Practical Wisdom of the Shamans of Ancient Mexico.* New York: HarperPerennial, 1998.

Castillo-Davis, C. I. "The Evolution of Noncoding DNA: How Much Junk, How Much Func?" *Trends in Genetics* 21, no. 10 (October 2005): 533–36.

Chen, Christopher, et al. "Geometric Control of Cell Life and Death." *Science* 276 (1997): 1425–28.

Cheney, Paul. "Chronic Fatigue, Mycotoxins, Abnormal Clotting and Other Notes." *Townsend Letter for Doctors and Patients.* tldp.com/issue/157-8/157pub.htm.

———. "New Insights into the Pathophysiology and Treatment of CFS." Presentation to the CFIDS and FMS Support Group of Dallas-Fort Worth, October 2001. Summary by Linda Sleffel. cfs-ireland.com/cheney2.htm.

Childre, Doc, and Howard Martin, with Donna Beech. *The Heartmath Solution.* San Francisco: HarperCollins, 1999.

Clarke, A. R., et al. "Thymocyte Apoptosis Induced by p53-Dependent and Independent Pathways." *Nature* 362 (April 1993): 849–52.

Cohen, Kenneth S. *The Way of Qigong.* New York: Ballentine, 1997.

Cohen, S., and S. L. Symeeds. *Social Support and Health.* New York: Academic Press, 1985.

Cohen, S., D. A. Tyrell, and A. P. Smith. "Psychological Stress and Susceptibility to the Common Cold." *New England Journal of Medicine* 325: (1991): 606–12.

Cohen, S., et al. "Human Relationships and Infectious Disease." *Journal of the American Medical Association* 277 (1997):1940–45.

Cong, Y. S., et al. "Human Telomerase and Its Regulation." *Microbiology and Molecular Biology Review* 66, no. 3 (September 2002): 407–25.

Cousins, Norman. *Anatomy of an Illness as Perceived By the Patient.* Toronto: Bantam, 1979.

———. *Head First: Biology of Hope and Healing Power of the Human Spirit.* New York: Penguin, 1989.

Crick, Francis. "The Genetic Code." In *What Mad Pursuit: A Personal View of Scientific Discovery*, 89–101. New York: Basic Books, 1988.

Cross, R. J., et al. "Hypothalamic-Immune Interactions." *Brain Research Journal* 196 (1980): 79–87.

Davidson, R. J., et al. "Alterations in Brain and Immune Function Produced by Mindfulness Meditation." *Psychosomatic Medicine* 65 (2003): 564–70.

de Duve, Christian. *Life Evolving: Molecules, Mind, and Meaning.* New York: Oxford University Press, 2002.

———. *Vital Dust: Life as a Cosmic Imperative*. New York: Basic Books, 1995.

De Volder, A. G., et al. "Auditory Triggered Mental Imagery of Shape Involves Visual Association Areas in Early Blind Humans." *Neuroimage* 14 (July 2001): 129–39.

Deamer, David W. "How Did It All Begin? The Self-Assembly of Organic Molecules and the Origin of Cellular Life Evolution:

Investigating the Evidence." *Paleontological Society Special Publication* 9 (1999).

Demarquay, G., P. Ryvlin, and J. P. Royet. "Olfaction and Neurological Diseases: A Review of the Literature." *Revue Neurologie (Paris)* 163 (2007): 155–67.

Denton, Michael. *Nature's Destiny: How the Laws of Biology Reveal Purpose in the Universe.* New York: Simon & Schuster, 1998.

Desai, A., and T. J. Mitchison. "Microtubule Polymerization Dynamics." *Annual Review of Cell Biology* 13 (1997): 83–117.

Dong, Seung Myung. "Detecting Colorectal Cancer in Stool with the Use of Multiple Genetic Targets." *Journal of the National Cancer Institute* 93, no. 11 (2001): 858–65.

Dossey, L. "The return of prayer." *Alternative Therapies in Health and Medicine* 3, no. 6 (1997):10–17, 113–20.

———. "How Healing Happens: Exploring the Nonlocal Gap." *Alternative Therapies in Health and Medicine* 8, no. 2 (2002): 12–16, 103–10.

———. *Meaning and Medicine.* New York: Bantam, 1991.

Einstein, Albert. "Ist die Trägheit eines Körpers von seinem Energieinhalt abhängig?" *Annalen der Physik* 18 (1905): 639–43.

Eisenberg, D. T. "An Evolutionary Review of Human Telomere Biology: The Thrifty Telomere Hypothesis and Notes on Potential Adaptive Paternal Effects." *American Journal of Human Biology* 23, no. 2 (2011): 149–67.

Eisenberg, David, with T. Wright. *Encounters with Qi: Exploring Chinese Medicine.* New York: Penguin, 1985.

Elgar, G., and T. Vavouri. "Tuning In to the Signals: Noncoding Sequence Conservation in Vertebrate Genomes." *Trends in Genetics* 24, no. 7 (July 2008): 344–52.

Epel, E. S., et al. "Dynamics of Telomerase Activity in Response to Acute Psychological Stress." *Brain, Behavior, and Immunity* 24, no. 4 (2010): 531–39.

Eremin, Oleg, et al. "Immuno-modulatory Effects of Relaxation Training and Guided Imagery in Women with Locally Advanced Breast Cancer Undergoing Multimodality Therapy: A Randomised Controlled Trial." *The Breast* 18, no. 1 (February 2009): 17–25.

Erickson, M. H. "Further Clinical Techniques of Hypnosis: Utilization Techniques." *American Journal of Clinical Hypnosis* 51, no. 4 (April 2009): 341–62.

———. "Special Inquiry with Aldous Huxley into the Nature and Character of Various States of Consciousness." *American Journal of Clinical Hypnosis* 8 (July 1965): 14–33.

Erickson, M. H., and E. L. Rossi. "Autohypnotic Experiences of Milton H. Erickson." *American Journal of Clinical Hypnosis* 20, no. 1 (July1977): 36–54.

Evans, N. D. "Substrate Stiffness Affects Early Differentiation Events in Embryonic Stem Cells." *European Cells and Materials* 18 (September 21, 2009): 1–14.

Fearon, Eric R. "Human Cancer Syndromes: Clues to the Origin and Nature of Cancer." *Science* 278, no. 5340 (November 1997): 1043–50.

Fell, J., N. Axmacher, and S. Haupt. "From Alpha to Gamma: Electrophysiological Correlates of Meditation-Related States of Consciousness." *Medical Hypotheses* 75, no. 2 (August 2010): 218–24.

Fenech, M. "Chromosomal Damage Rate, Aging and Diet." *Annals of the New York Academy of Sciences* 854 (1998): 23–36.

Florez, H., et al. "C-Reactive Protein Is Elevated in Obese Patients with the Metabolic Syndrome." *Diabetes Research and Clinical Practice* 71, no. 1 (2006): 92–100.

Fontani, G., et al. "Effect of Mental Imagery on the Development of Skilled Motor Actions." *Perceptual and Motor Skills* 105, no. 3, pt. 1 (December 2007): 803–26.

Francis, M., and J. W. Pennebaker. "Putting Stress into Words: The Impact of Writing on Physiological, Absentee, and Self-Reported Emotional Well-Being Measures." *American Journal of Health Promotion* 6 (1992): 280–87.

Frankenhaeuser, M., et al. "Sex Differences in Psychoneuroendocrine Reactions to Examination Stress." *Psychosomatic Medicine* 40, no. 4 (1978): 334–43.

Fuller, Buckminster. "Conceptuality of Fundamental Structures." In *Structure in Art and in Science*, edited by G. Kepes, 66–88. New York: Braziller, 1965.

———. "Tensegrity." *Portfolio Artnews Annual* 4 (1961): 112–27.

Furlow, F. Bryant. "The Smell of Love: How Women Rate the Sexiness and Pleasantness of a Man's Body Odor Hinges on How Much of Their Genetic Profile Is Shared." *Psychology Today* 29 (1996): 38.

Gardner, Russell, and Gerald A. Cory. *The Evolutionary Neuroethology of Paul MacLean: Convergences and Frontiers.* New York: Praeger, 2002.

Garfield, C. A., with H. Z. Bennet. *Peak Performance: Mental Training Techniques from the World's Greatest Athletes.* New York: Warner Books, 1984.

GDB Human Genome Database Hosted by RTI International [online]. North Carolina. Available from: gdb.org/gdbreports/CountGeneByChromosome.html

"Genes and Chromosomes." Centre for Genetics Education. Internet: genetics.edu.au/factsheet/fs1.

German, J. Bruce, et al. "Olfaction: Where Nutrition, Memory and Immunity Intersect." In *Flavors and Fragrances*, edited by Ralf Berger, 25–32. Berlin: Springer-Verlag, 2007.

Ghanta, V., et al. "Neural and Environmental Influences on Neoplasia and Conditioning of NK Activity." *Journal of Immunology* 135 (1985): 848–52.

Goh, A. M., C. R. Coffill, and D. P. Lane. "The Role of Mutant p53 in Human Cancer." *Journal of Pathology* 223, no. 2 (January 2011): 116–26.

Gordon, Ilanit, et al. "Oxytocin and the Development of Parenting in Humans." *Biological Psychiatry* 68 (2010): 377–82.

Gregory, T. R. "Genome Size Evolution in Animals." In *The Evolution of the Genome*, edited by T. R. Gregory, 4–71. San Diego, CA: Elsevier, 2005.

Håberg, S. E., et al. "Folic Acid Supplements in Pregnancy and Early Childhood Respiratory Health." *Archives of Disease in Childhood* 94 (2009): 180–84.

Haffner, S. M. "The Metabolic Syndrome: Inflammation, Diabetes Mellitus, and Cardiovascular Disease." *American Journal of Cardiology* 97 (2006): 3A–11A.

Hainaut, P., and M. Hollstein. "p53 and Human Cancer: The First Ten Thousand Mutations." *Advances in Cancer Research* 77 (2000): 82–137.

Hameroff, Stuart R. "Ch'i: A Neural Hologram? Microtubules, Bioholography, and Acupuncture." *American Journal of Chinese Medicine* 2, no. 2 (1974): 163–70.

———. "The Entwined Mysteries of Anesthesia and Consciousness: Is There a Common Underlying Mechanism?" *Anesthesiology* 105 (2006): 400–12.

Hameroff, Stuart R., and Roger Penrose. "Conscious Events as Orchestrated Spacetime Selections." *Journal of Consciousness Studies* 3, no. 1 (1996): 36–53.

———. "Orchestrated Reduction of Quantum Coherence in Brain Microtubules: A Model for Consciousness?" In *Toward a Science of Consciousness: The First Tucson Discussions and Debates*, edited by S. R. Hameroff, A. W. Kaszniak, and A. C. Scott, 507–40. Cambridge, MA: MIT Press, 1996.

Hameroff, Stuart, et al. "Conduction Pathways in Microtubules, Biological Quantum Computation, Consciousness." *Biosystems* 64, nos. 1–3 (2002): 149–68.

Hardy, K. "Apoptosis in the Human Embryo." *Reviews of Reproduction* 4 (1999): 125–34.

Harner, Michael. *The Way of the Shaman: A Guide to Power and Healing*. New York: Bantam, 1980.

Harold, Franklin M., "Molecules into Cells: Specifying Spatial Architecture." *Microbiology and Molecular Biology Reviews* 69 (2005): 544–64.

Harris, W. S., et al. "A Randomized, Controlled Trial of the Effects of Remote, Intercessory Prayer on Outcomes in Patients Admitted to the Coronary Care Unit." *Archives of Internal Medicine* 159 (October 1999): 2273–8.

Hart, Mickey, with Jay Stevens. *Drumming at the Edge of Magic: A Journey into the Spirit of Percussion.* San Francisco: HarperSanFrancisco, 1990.

Hassett, James. "The Sweat Gland." In *A Primer of Psychophysiology*, 32–46. San Francisco: W. H. Freeman, 1978.

Hazum, E., K. J. Chang, and P. Cuatrecasas. "Specific Non-opiate Receptors for β-Endorphins on Human Lymphocytes." *Science* 205 (1970): 1033–35.

Heinrichs, Markus, et al. "Social Support and Oxytocin Interact to Suppress Cortisol and Subjective Responses to Psychosocial Stress." *Biological Psychiatry* 54 (2003): 1389–98.

Hoffman-Goetz, Laurie, and Bente Klarlund Pedersen. "Exercise and the Immune System: A Model of the Stress Response?" *Immunology Today* 15 (1994): 382–87.

Holden, C. "Paul MacLean and the Triune Brain." *Science* 204, no. 4397 (June 8, 1979): 1066–68.

Hooper, J., and D. Teresi. *The Three-Pound Universe.* London: Macmillan, 1986.

Horgan, Bonnie. "Consciousness, Microtubules and the Quantum World: Interview with Stuart Hameroff, MD." *Alternative Therapies* 3 (May 1997): 70–79.

House, J. S., K. R. Landis, and D. Umberson. "Social Relationships and Health." *Science* 241 (1988): 540–45.

Huang, Sui, and Donald E. Ingber. "Cell Tension, Matrix Mechanics, and Cancer Development." *Cancer Cell* 8, no. 3 (September 2005): 175–76.

Iggo, R. "Increased Expression of Mutant Forms of p53 Oncogene in Primary Lung Cancer." *Lancet* 335, no. 8691 (March 1990): 675–79.

Ingber, Donald E. "The Architecture of Life." *Scientific American* 278, no. 1 (January 1998): 47–57.

———. "Cellular Tensegrity: Defining the New Rules of Biological Design that Govern the Cytoskeleton." *Journal of Cell Science* 104 (1993): 613–27.

———. "Tensegrity I. Cell Structure and Hierarchical Systems Biology." *Journal of Cell Science* 116 (2003): 1157–73.

———. "Tensegrity: The Architectural Basis of Cellular Mechanotransduction." *Annual Review of Physiology* 59 (1997): 575–99.

———. "Tensegrity-Based Mechanosensing from Macro to Micro." *Progress in Biophysics and Molecular Biology* 97 (2008): 163–79.

Irwin, M., et al. "Partial Sleep Deprivation Reduces Natural Killer Cell Activity in Humans." *Psychosomatic Medicine* 56 (1994): 493–98.

Jung, Carl. *Collected Works.* Vol. 11, *Psychology and Religion: East and West.* Princeton, NJ: Princeton University Press, 1969.

———. *Man and His Symbols.* New York: Doubleday, 1964.

———. *Memories, Dreams and Reflections.* New York: Alfred A. Knopf, 1961.

Kabat-Zinn, Jon. *Wherever You Go, There You Are: Mindfulness Meditation in Everyday Life.* New York: Hyperion, 1994.

Kadouri, L., et al. "Cancer Risks in Carriers of the BRCA1/2 Ashkenazi Founder Mutations." *Journal of Medical Genetics* 44, no. 7 (2007): 467–71.

Kenfield, S. A., et al. "Smoking and Smoking Cessation in Relation to Mortality in Women." *Journal of the American Medical Association* 299, no. 17 (May 2008): 2037–47.

Khajavinia, A., and W. Makalowski. "What Is Junk DNA, and What Is It Worth?" *Scientific American* 296, no. 5 (May 2007): 104.

Kiecolt-Glaser, J., et al. "Marital Quality, Marital Disruption, and Immune Function. *Psychosomatic Medicine* 49 (1987): 13–34.

———. "Psychosocial Modifiers of Immunocompetence in Medical Students." *Psychosomatic Medicine* 46 (1984): 7–14.

———. "Slowing of Wound Healing by Stress." *Lancet* 346 (1995): 1194–96.

Koshland, D. E. "Molecule of the Year." *Science* 262, no. 5142 (December 1993): 1953.

Kosslyn, Stephen M., et al. "Topographic Representations of Mental Images in Primary Visual Cortex." *Nature* 378 (1995): 496–98.

———. "Two Types of Image Generation: Evidence from PET." *Cognitive, Affective & Behavioral Neuroscience* 5 (2005): 41–53.

Kroeger, M. "Oxytocin: Key Hormone in Sexual Intercourse, Parturition, and Lactation." *The Birth Gazette* 13 (1996): 28–30.

la Fougère, C., et al. "Real Versus Imagined Locomotion: A [18F]-FDG PET-fMRI Comparison." *Neuroimage* 50, no. 4 (May 1, 2010): 1589–98.

Lafreniere, D., and N. Mann. "Anosmia: Loss of Smell in the Elderly." *Otolaryngologic Clinics of North America* 42 (2009): 123–31.

Laurance, Jeremy. "Why Women Can't Sniff Out Mr. Right When They Take the Pill." *The Independent Health News*, August 2008.

Lauterwasser, Alexander. *Water Sound Images.* New Market, NH: Macromedia, 2006.

Lee, D. A., et al. "Stem Cell Mechanobiology." *Journal of Cellular Biochemistry* 112, no. 1 (January 2011): 1–9.

Leonard, George. *The Silent Pulse.* New York: Bantam, 1981.

Levental, Kandice R., et al. "Matrix Crosslinking Forces Tumor Progression by Enhancing Integrin Signaling." *Cell* 139, no. 5 (November 25, 2009): 891–906.

Li, G., and V. C. Ho. "p53-Dependent DNA Repair and Apoptosis Respond Differently to High- and Low-Dose Ultraviolet

Radiation." *The British Journal of Dermatology* 139, no. 1 (July 1998): 3–10.

Lipton, Bruce H. *The Biology of Belief: Unleashing the Power of Consciousness, Matter and Miracles.* Santa Rosa, CA: Mountain of Love/Elite Books, 2005.

Lloyd, A, O. Brett, and K. Wesnes. "Coherence Training in Children with Attention-Deficit Hyperactivity Disorder: Cognitive Functions and Behavioral Changes." *Alternative Therapies in Health and Medicine* 16, no. 4 (July/August 2010): 34–42.

Locke, S., and L. Kraus. "Modulation of Natural Killer Cell Activity by Life Stress and Coping Ability." In *Biological Mediators of Behavior and Disease: Neoplasia*, edited by S. Levy, 3–28. New York: Elsevier, 1982.

Loeb, Lawrence A, K. Loeb, and J. Anderson. "Multiple Mutations and Cancer." *Proceedings of the National Academy of Sciences of the United States of America* 100, no. 3 (February 4, 2003): 776–81.

Lusis, A. J. "Atherosclerosis." *Nature* 407, no. 6801 (2000): 233–41.

Lyles, Jeanne N., et al. "Efficacy of Relaxation Training and Guided Imagery in Reducing the Aversiveness of Cancer Chemotherapy." *Journal of Consulting and Clinical Psychology* 50, no. 4 (August 1982): 509–24.

Macfarlane, A. "Olfaction in the Development of Social Preferences in the Human Neonate." *Ciba Foundation Symposium* 33 (1975): 103–17.

MacLean, Paul. "The Triune Brain in Conflict." *Psychotherapy and Psychosomatics* 28, nos. 1–4 (1977): 207–20.

Mahlberg, Arden. "Therapeutic Healing with Sound." In *Music and Miracles*, compiled by Don Campbell, 219–29. Wheaton, IL: Quest Books, 1992.

Margulis, Lynn, and Dorion Sagan. *Microcosmos: Four Billion Years of Microbial Evolution.* Berkeley, CA: University of California Press, 1986.

Marks, D. F. "New Directions for Mental Imagery Research." *Journal of Mental Imagery* 19 (1995): 153–67.

Maslinski, W., E. Grabczewska, and J. Ryzewski. "Acetylcholine Receptors of Rat Lymphocytes." *Biochimica et Biophysica Acta* 663 (1980): 269–73.

May, P., and E. May. "Twenty Years of p53 Research: Structural and Functional Aspects of the p53 Protein." *Oncogene* 18 (1999): 7621–36.

Mayo Clinic. "Loss of Smell (Anosmia)." February 8, 2011. mayoclinic.com/health/loss-of-smell/MY00408/DSECTION=causes.

McCraty, R. "Coherence: Bridging Personal, Social, and Global Health." *Alternative Therapies in Health and Medicine* 16, no. 4 (July/August 2010): 10–24.

Ming, Ye, et al. "Stress-Reducing Practice of Qigong Improved DNA Repair in Cancer Patients." Shanghai Qigong Institute, 2nd World Conference on Academic Exchange of Medical Qigong, 1993.

Misteli, Tom. "The Inner Life of the Genome." *Scientific American* 304, no. 2 (February 2011): 66–73.

Misteli, Tom, and David Spector, eds. *The Nucleus.* Cold Spring Harbor, NY: Cold Spring Harbor Laboratory Press, 2010.

Molinoff, P. B., and J. Axelrod. "Biochemistry of Catecholamines." *Annual Review of Biochemistry* 40 (1971): 465–500.

Moons, W. G., N. I. Eisenberger, and S. E. Taylor. "Anger and Fear Responses to Stress Have Different Biological Profiles." *Brain, Behavior, and Immunity* 24 (2010): 215–19.

Naber, Fabienne, et al. "Intranasal Oxytocin Increases Fathers' Observed Responsiveness during Play with Their Children: A Double-Blind Within-Subject Experiment." *Psychoneuroendocrinology* 35 (2010): 1583–86.

Narby, Jeremy. *The Cosmic Serpent: DNA and the Origins of Knowledge.* New York: Tarcher/Putnam, 1998.

National Center for Complementary and Alternative Medicine. National Institutes of Health. nccam.nih.gov.

Newman, John D., and James C. Harris. "The Scientific Contributions of Paul D. MacLean (1913–2007)." *Journal of Nervous and Mental Disease* 197, no. 1 (2009): 1–2.

Newman, M. G. "Can an Immune Response Be Conditioned?" *Journal of the National Cancer Institute* 82 (1990): 1543–45.

Ng, Mei Rosa, and Joan S. Brugge. "A Stiff Blow from the Stroma: Collagen Crosslinking Drives Tumor Progression." *Cancer Cell* 16, no. 8 (2009): 455–57.

Nobelprize.org. "The Cell and its Organelles." August 14, 2011. nobelprize.org/educational/medicine/cell/.

Oh, B., et al. "A Critical Review of the Effects of Medical Qigong on Quality of Life, Immune Function, and Survival in Cancer Patients." *Integrative Cancer Therapies* 11 (June 28, 2011): 101–10.

———. "Impact of Medical Qigong on Quality of Life, Fatigue, Mood and Inflammation in Cancer Patients: A Randomized Controlled Trial." *Annals of Oncology* 21, no. 3 (March 2010): 608–14.

Ornish, Dean, et al. "Changes in Prostate Gene Expression in Men Undergoing an Intensive Nutrition and Lifestyle Intervention." *Proceedings of the National Academy of Sciences of the United States of America* 105 (June 2008): 8369–74.

Palomaki, A., et al. "Ubiquinone Supplementation during Lovastatin Treatment: Effect on LDL Oxidation Ex Vivo." *Journal of Lipid Research* 39 (1998): 1430–37.

Parham, P., and T. Ohta. "Population Biology of Antigen Presentation by MHC Class 1 Molecules." *Science* 272 (1996): 67–74.

Paszek, Matthew J., et al. "Tensional Homeostasis and the Malignant Phenotype." *Cancer Cell* 8 (September 2005): 241–54.

Pearsall, Paul. *The Heart's Code: Tapping the Wisdom and Power of Our Heart Energy.* New York: Broadway Books, 1998.

Pearsall, Paul, Gary E. Schwartz, and Linda G. Russek. "Organ Transplants and Cellular Memories." *Nexus Magazine* 12, no. 3 (April/May 2005).

Pennebaker, James. *Opening Up: The Healing Power of Expressing Emotions.* New York: Guilford Press, 1991.

Pennebaker, James, et al. "Confronting Traumatic Experiences and Health among Holocaust Survivors." *Advances* 6 (1989): 14–17.

Pennebaker, James, Kiecolt-Glaser, J. K., and R. Glaser. "Disclosure of Traumas and Immune Function: Health Implications." *Journal of Consulting and Clinical Psychology* 56 (1988): 239–45.

Pennisi, Elizabeth. "DNA Study Forces Rethink of What It Means to Be a Gene." *Science* 316, no. 5831 (2007): 1556–7.

Penrose, Roger. *The Emperor's New Mind.* Oxford, UK: Oxford University Press, 1989.

Pert, C. B., et al. "Neuropeptides and Their Receptors: A Psychosomatic Network." *Journal of Immunology* 135 (1985): 118–22.

Pfeifer, G. P., et al. "Tobacco Smoke Carcinogens, DNA Damage and p53 Mutations in Smoking-Associated Cancers." *Oncogene* 21 (2002): 7435–51.

Pribram, Karl. *Languages of the Brain: Experimental Paradoxes and Principles in Neuropsychology.* Englewood Cliffs, NJ: Prentice-Hall, 1971.

Pribram, Karl, ed. *Rethinking Neural Networks: Quantum Fields and Biological Data.* Hillsdale, NJ: Erlbaum, 1993.

Pribram, Karl, and Donald Broadbent, eds. *Biology of Memory.* New York: Academic Press, 1970.

Pribram, Karl, et al. "The Holographic Hypothesis of Memory Structures in Brain Function and Perception." In *Contemporary Developments in Mathematical Psychology*, vol. II, edited by R. C. Atkinson et al. San Francisco: W.H. Freeman, 1974.

Purce, Jill. *The Mystic Spiral: Journey of the Soul.* London: Thames & Hudson, 1980.

Rapaport, M. H., P. Schettler, and C. Bresee. "A Preliminary Study of the Effects of a Single Session of Swedish Massage on Hypothalamic-Pituitary-Adrenal and Immune Function in Normal Individuals." *Journal of Alternative and Complementary Medicine* 16 (2010): 1079–88.

Reichlin, Seymour. "Neuroendocrine-Immune Interactions." *New England Journal of Medicine* 329 (1993): 1246–53.

Rensberger, Boyce. *Life Itself: Exploring the Realm of the Living Cell.* New York: Oxford University Press, 1996.

Roitt, Ivan, David Male, and Jonathan Brostoff. *Immunology.* 4th ed. St. Louis, MO: Mosby Year Book, 1996.

Rood, Y. R., et al. "The Effects of Stress and Relaxation on the In Vitro Immune Response in Man: A Meta-Analytic Study." *Journal of Behavioral Medicine* 16 (1993): 163–81.

Rose, N. R. "Mechanisms of Autoimmunity." *Seminars in Liver Disease* 22 (2002): 387–94.

Rosen, Sidney, ed. *My Voice Will Go with You: The Teaching Tales of Milton H. Erickson.* New York: Norton, 1991.

Rossi, Ernest. *The Psychobiology of Mind-Body Healing: New Concepts of Therapeutic Hypnosis.* Revised edition. New York: Norton, 1993.

Sachs, L. "The Adventures of a Biologist: Prenatal Diagnosis, Hematopoiesis, Leukemia, Carcinogenesis, and Tumor Suppression." *Advances in Cancer Research* 66 (1995): 1–40.

———. "The Control of Hematopoiesis and Leukemia: From Basic Biology to the Clinic." *Proceedings of the National Academy of Sciences of the United States of America* 93 (1996): 4742–49.

Schleifer, S., et al. "Lymphocyte Function in Major Depressive Disorder." *Archives of General Psychiatry* 41 (1984): 484–86.

Schneider, Michael S. *A Beginner's Guide to Constructing the Universe: The Mathematical Archetypes of Nature, Art, and Science.* New York: HarperPerennial, 1994.

Selivanova, Galina. "p53: Fighting Cancer." *Current Cancer Drug Targets* 4, no. 5 (August 2004): 385–402.

Seyle, H. *The Stress of Life.* New York: McGraw-Hill, 1956.

Sheikh, Anees, ed. *Imagination and Healing.* Farmingdale, NY: Baywood Publishing, 1984.

Sheldrake, Rupert. "Morphic Resonance and Morphic Fields—An Introduction." February 2005. sheldrake.org/Articles&Papers/papers/morphic/morphic_intro.html.

———. *The Rebirth of Nature: The Greening of Science and God.* New York: Bantam, 1991.

Shlain, Leonard. *Art and Physics.* New York: HarperPerennial, 1991.

Siegel, G. J., et al., ed. *Basic Neurochemistry: Molecular, Cellular and Medical Aspects.* 6th edition. Philadelphia: Lippincott-Raven, 1999.

Singh, R. B., et al. "Randomized, Double-Blind Placebo-Controlled Trial of Coenzyme Q10 in Patients with Acute Myocardial Infarction." *Cardiovascular Drugs and Therapy* 12 (1998): 347–53.

Slagter, H. A., R. J. Davidson, and A. Lutz, "Mental Training as a Tool in the Neuroscientific Study of Brain and Cognitive Plasticity." *Frontiers in Human Neuroscience* 5 (February 2011): 1–12.

Smyth, J. K., et al. "Effects of Writing About Stressful Experiences on Symptom Reduction in Patients with Asthma or Rheumatoid Arthritis." *Journal of the American Medical Association* 281 (1999): 1304–09.

Sobel, D. "Rethinking Medicine: Improving Health Outcomes with Cost-Effective Psychosocial Interventions." *Psychosomatic Medicine* 57 (1995): 234–37.

Spiegel, David. "Imagery and Hypnosis in the Treatment of Cancer Patients." *Oncology* 11, no. 8 (1997): 1–15.

Spiegel, D., et al. "Psychological Support for Cancer Patients." *Lancet* 2 (1989): 1447–49.

Stanford University. "The Equivalence of Mass and Energy." Stanford Encyclopedia of Philosophy online. First published September 12, 2001; substantive revision June 23, 2010. plato.stanford.edu/entries/equivME/.

Stapleton, M. P. "Sir James Black and Propranolol: The Role of the Basic Sciences in the History of Cardiovascular Pharmacology." *Texas Heart Institute Journal* 24 (1997): 336–42.

Stoll, G., and M. Bendszus. "Inflammation and Atherosclerosis: Novel Insights into Plaque Formation and Destabilization." *Stroke* 37 (2006): 1923–32.

Strous, Raul D., and Yehuda Shoenfeld. "To Smell the Immune System: Olfaction, Autoimmunity and Brain Involvement." *Autoimmunity Reviews* 6 (2006): 54–60.

Stryer, Lubert. *Biochemistry.* 5th ed. New York: W. H. Freeman, 2002.

Sullivan, Regina M., and Paul Toubas. "Clinical Usefulness of Maternal Odor in Newborns: Soothing and Feeding Preparatory Responses." *Biology of the Neonate* 74 (1998): 402–8.

Syrjala, Karen L., et al. "Relaxation and Imagery and Cognitive-Behavioral Training Reduce Pain during Cancer Treatment: A Controlled Clinical Trial." *Pain* 63, no. 2 (November 1995): 189–98.

Talbot, Michael. *The Holographic Universe.* New York: HarperCollins, 1991.

Taylor, S. E. *The Tending Instinct: Women, Men, and the Biology of Relationships.* New York: Holt, 2003.

Taylor, S. E., S. Saphire-Bernstein, and T. E. Seeman. "Are Plasma Oxytocin in Women and Plasma Vasopressin in Men Biomarkers of Distressed Pair Bond Relationships?" *Psychological Science* 21 (2010): 3–7.

Teilhard de Chardin, Pierre. *The Phenomenon of Man.* Translation by Bernard Wall. New York: Harper and Row, 1961.

Thayer, Robert E. "Energy, Tiredness, and Tension Effect of a Sugar Snack Versus Moderate Exercise." *Journal of Personality and Social Psychology* 52 (1987): 119–25.

———. *The Origin of Everyday Moods: Managing Energy, Tension, and Stress.* New York: Oxford University Press, 1996.

Thomas, Lewis. *The Lives of a Cell: Notes of a Biology Watcher.* New York: Bantam, 1974.

Thorsby, E., and B. A. Lie. "Certain HLA Patterns Signify the Likelihood of Developing an Autoimmune Disease." *Transplant Immunology* 14 (2005):175–82.

Thurman, Robert. *The Jewel Tree of Tibet.* New York: Free Press, 2005.

Tobias, L. "A Briefing Report on Autoimmune Diseases and AARDA: Past, Present, and Future." Detroit, MI: American Autoimmune Related Diseases Association, 2010.

Turner, D. D. "Just Another Drug? A Philosophical Assessment of Randomised Controlled Studies on Intercessory Prayer." *Journal of Medical Ethics* 32 (2006): 487–90.

Uchino, B. N., J. T. Cacioppo, and J. K. Kiecolt-Glaser. "The Relationship between Social Support and Physiological Processes: A Review with Emphasis on Underlying Mechanisms." *Psychological Bulletin* 119 (1996): 488–531.

United States National Institutes of Health. "The p53 Tumor Suppressor Protein." In *Genes and Disease.* Bethesda, MD: National Center for Biotechnology Information, 1998. ncbi.nlm.nih.gov/books/NBK22183/.

Van Over, Raymond. *Sun Songs: Creation Myths from Around the World.* New York: Signet Books, 1980.

Vaughan, Christopher. *How Life Begins: The Science of Life in the Womb.* New York: Dell, 1997.

Ventura, A., et al. "Restoration of p53 Function Leads to Tumour Regression in Vivo." *Nature* 445, no. 7128 (2007): 661–65.

Vincent, C., and J. P. Revillard. "Characterization of Molecules Bearing HLA Determinants in Serum and Urine." *Transplantation Proceedings* 11 (1979): 1301–2.

Vogelstein, Bert, and Kenneth W. Kinzler. "The Multistep Nature of Cancer." *Trends in Genetics* 9, no. 4 (April 1993): 138–41.

Waterland, R. A., and R. L. Jirtle. "Transposable Elements: Targets for Early Nutritional Effects on Epigenetic Gene Regulation." *Molecular and Cell Biology* 23 (2003): 5293–300.

Watson, J. D., et al. *Molecular Biology of the Gene.* San Francisco: Pearson/Benjamin Cummings, 2008.

Watters, Ethan. "DNA Is Not Destiny: The New Science of Epigenetics Rewrites the Rules of Disease, Heredity, and Identity." *Discover*, November 2006, 33–37, 75.

Wedekind, Claus, et al. "MHC-Dependent Mate Preferences in Humans." *Proceedings of the Royal Society of London* 260 (1995): 245–49.

Weinshenker, B. G. "Natural History of Multiple Sclerosis." *Annals of Neurology* 36 (1994): S6–S11.

White, Ray. "Inherited Cancer Genes." *Current Opinion in Genetics & Development* 2, no. 1 (February 1992): 53–57.

Wickramasekera, I. "A Conditioned Response Model of the Placebo Effect." In *Placebo: Theory, Research and Mechanisms*, edited by L. White et al., 255–287. New York: Guilford Press, 1985.

Winkelman, Michael. *Shamanism: The Neural Ecology of Consciousness and Healing.* Westport, CT: Bergin & Garvey, 2000.

Wolf, S. L., et al. "Reducing Frailty and Falls in Older Persons: An Investigation of Tai Chi and Computerized Balance Training. Atlanta FICSIT Group. Frailty and Injuries: Cooperative Studies of Intervention Techniques." *Journal of the American Geriatrics Society* 44, no. 5 (May 1996): 489–97.

Wu, Ge. "Evaluation of the Effectiveness of Tai Chi for Improving Balance and Preventing Falls in the Older Population—A Review." *Journal of the American Geriatrics Society* 50 (April 2002): 746–54.

Yan, Johnson F. *DNA and the I Ching: The Tao of Life.* Berkeley, CA: North Atlantic Books, 1991.

Zeidan, F., et al. "Mindfulness Meditation Improves Cognition: Evidence of Brief Mental Training." *Consciousness and Cognition* 19, no. 2 (June 2010): 597–605.

Zeisel, Steven H. "Importance of Methyl Donors During Reproduction." *American Journal of Clinical Nutrition* 89 (February 2009): S673–S677.

Suggested Reading

Body-Mind Essentials

Pearsall, Paul. *The Heart's Code: Tapping the Wisdom and Power of Our Heart Energy.* New York: Broadway Books, 1998.

Seligman, Martin. *What You Can Change and What You Can't.* New York: Ballentine, 1995.

Sternberg, Esther. *The Balance Within: The Science Connecting Health and Emotions.* New York: W.H. Freeman, 2000.

DNA and Genes

DNA Learning Center Blog. "Blackburn, Greider and Szostak Share Nobel for Telomeres." blogs.dnalc.org/dnaftb/2009/10/05/blackburn-greider-and-szostak-share-nobel-for-telomeres=2/.

Lipton, Bruce H. *The Biology of Belief: Unleashing the Power of Consciousness, Matter and Miracles.* Santa Rosa, CA: Mountain of Love/Elite Books, 2005.

Energy, Stress, and Relaxation

Becker, Robert. *The Body Electric: Electromagnetism and the Foundation of Life.* New York: William Morrow, 1985.

Benson, Herbert, with Miriam Z. Klipper. *The Relaxation Response.* New York: William Morrow, 1975.

Gerber, Richard. *Vibrational Medicine.* Santa Fe, NM: Bear & Company, 1988.

Hameroff, Stuart. Quantum Consciousness. quantumconsciousness.org/.

Sapolsky, Robert. *Why Zebras Don't Get Ulcers: A Guide to Stress, Stress-Related Diseases, and Coping.* New York: W. H. Freeman, 1998.

Wolf, Fred Alan. *The Body Quantum.* New York: Macmillan, 1986.

General Biology

de Duve, Christian. *Life Evolving: Molecules, Mind, and Meaning.* New York: Oxford University Press, 2002.

———. *Vital Dust: Life as a Cosmic Imperative.* New York: Basic Books, 1995.

Denton, Michael. *Nature's Destiny: How the Laws of Biology Reveal Purpose in the Universe.* New York: Simon & Schuster, 1998.

Hoagland, Mahlon, and Bert Dodson. *The Way Life Works.* New York: Times Books, 1995.

Rensberger, Boyce. *Life Itself: Exploring the Realm of the Living Cell.* New York: Oxford University Press, 1996.

Thomas, Lewis. *The Lives of a Cell: Notes of a Biology Watcher.* New York: Bantam, 1974.

Vaughan, Christopher. *How Life Begins: The Science of Life in the Womb.* New York: Dell, 1997.

Immune System

Nobelprize.org. "Blood Typing." September 16, 2011. nobelprize.org/educational/medicine/landsteiner/.

Nobelprize.org. "The Immune System—In More Detail." September 16, 2011. nobelprize.org/educational/medicine/immunity/immune-detail.html.

Indigenous and Shamanic Wisdom

Narby, Jeremy. *The Cosmic Serpent: DNA and the Origins of Knowledge.* New York: Tarcher/Putnam, 1998.

Pinkson, Tom. *The Shamanic Wisdom of the Huichol: Medicine Teachings for Modern Times.* Rochester, VT: Destiny Books, 2010.

Suzuki, David, and Peter Knudtsen. *Wisdom of the Elders: Honoring Sacred Native Visions of Nature.* New York: Bantam, 1992.

Winkler, Gershon. *Magic of the Ordinary: Recovering the Shamanic in Judaism.* Berkeley, CA: North Atlantic Books, 2003.

Sound Healing

David, William. *The Harmonics of Sound, Color and Vibration: A System for Self-Awareness and Soul Evolution.* Marina del Rey, CA: DeVorss, 1980.

Dewhurst-Maddock, Olivia. *The Book of Sound Therapy.* New York: Simon & Schuster, 1993.

Goldman, Jonathan, *Healing Sounds: The Power of Harmonics.* Rockport, MA: Element, 1992.

Mahlberg, Arden. The Integral Psychology Center. integralpsychology.com.

The Senses and Imagery

History of Hypnosis. historyofhypnosis.org/milton-erickson/.

Pribram, Karl. Studies of the Brain. karlhpribram.net/.

Samuels, Michael, and Nancy Samuels. *Seeing with the Mind's Eye: The History, Techniques and Uses of Visualization.* New York: Random House, 1975.

Symbolism

Joseph Campbell Foundation. jcf.org/new/index.php.

Nozedar, Adele. *The Illustrated Signs & Symbols Sourcebook: An A to Z Compendium of over 1000 Designs.* New York: Metro Books, 2010.

Schneider, Michael S. *A Beginner's Guide to Constructing the Universe: The Mathematical Archetypes of Nature, Art, and Science.* New York: HarperPerennial, 1994.

Index

Italic page numbers refer to illustrations.

D

E

S

About the Author

Sondra Barrett earned her degree in biochemistry from the University of Illinois Medical School and completed postdoctoral training in immunology/hematology at the University of California Medical School, where she remained on the faculty for more than a decade. She developed both basic cancer research and supportive care programs while working with children with life-threatening illnesses, and began studying expressive arts, energy practices, and shamanism.

Sondra's photomicrographs have won awards from both Nikon and Olympus and have appeared in *Scientific American* magazine, Lawrence Hall of Science, Napa Valley Museum, and numerous other venues and international publications. Her thousands of presentations using this art form are touchstones for learning about health, science, sacred symbology, wine, and the senses. Her workshops and lectures have received acclaim internationally.

Sondra lives in Northern California. Her first book, *Wine's Hidden Beauty,* explores the inner world of wine and our senses and is available on her website and at bookstores, museums, and wineries. Visit sondrabarrett.com.

About Sounds True

Sounds True is a multimedia publisher whose mission is to inspire and support personal transformation and spiritual awakening. Founded in 1985 and located in Boulder, Colorado, we work with many of the leading spiritual teachers, thinkers, healers, and visionary artists of our time. We strive with every title to preserve the essential "living wisdom" of the author or artist. It is our goal to create products that not only provide information to a reader or listener, but that also embody the quality of a wisdom transmission.

For those seeking genuine transformation, Sounds True is your trusted partner. At SoundsTrue.com you will find a wealth of free resources to support your journey, including exclusive weekly audio interviews, free downloads, interactive learning tools, and other special savings on all our titles.

To listen to a podcast interview with Sounds True publisher Tami Simon and author Sondra Barrett, please visit SoundsTrue.com/bonus/SecretsofYourCells.